What I Stand On

What I Stand On

Practical Advice and Cantankerous Musings
from a Pioneering Organic Farmer

Bill Dow

Edited by Fred Broadwell
Preface by Daryl Walker
Foreword by Isaiah Allen
Photographs by Debbie Roos

ORGANIC LEAF PRESS | DURHAM, NC

First edition
Library of Congress Control Number 2015920741
ISBN 978-0-9970434-0-2

Design by Chris Crochetière, BW&A Books, Inc.
Unless otherwise noted, all photos are courtesy of Debbie Roos.

Dedicated to all the young (and not so young) farmers
who are "coming along"
and to those who assist them.

"What I stand for is what I stand on."
—Wendell Berry, "A Part"

Contents

Preface

Who was Bill Dow? He was a physician, a farmer, and a teacher. He was also a pioneer, historian, activist, advocate, mentor, friend, and skeptic. And for ten years, until his death in 2012, he was my beloved.

Bill possessed a wry sense of humor, a courageous spirit, and self-confidence tempered with a solid dose of humility. He rarely spoke of his own accomplishments and never in detail. Through farming, medicine, history and conservation, Bill was a dedicated teacher, a man who enjoyed watching students of all ages learn. He loved books and read voraciously to support what he taught.

Bill pursued his family's history doggedly and taught others how to trace their roots. His many winters spent researching in England and Scotland led him to the Scottish Agricultural Revolution and resulting diaspora, about which he hoped to write. Though his copious notes never became the book he dreamed of writing, they are available at the Madison, Indiana, Historical Society.

As a visionary, persevering physician, Bill and his Vanderbilt team set new standards for delivering health care in the 1970s, through exceptional community organizing in small, isolated Appalachian communities. The health project he helped create in East Tennessee, the Appalachian Student Health Coalition, is known to be one of the most innovative of its time.

Yet Bill's main passion was farming—he thrived on the fresh air, the dirt, the smells, the birds, the organic methods, and the beautiful crops. He took pride in the produce he grew and in the farm he made, which he loved to show to visitors. Even so, farming also caused him great angst—the weather, the wrong seed orders, the lost crops, the need to find helpers, the expenses, the limited income. From the early 1990s, when he was diagnosed with heart disease, Bill worried that he might have to be hospitalized for surgery with no health insurance— too costly, and Medicare a long way off.

Bill valued doing things the old-fashioned way, and his sense of integrity was profound. Contracts and legal documents were not for him; a handshake and his word were his gentleman's agreement. He treated others with care and respect, and he was especially sensitive to those he employed—be they student interns or Latino farmhands, central to his work and like family.

He clung to old pathways, many admirable and worthy, but his resistance to certain new things, like emailing chefs his weekly produce list, was absolutely unshakeable and rigid. When we first met, Bill assured me, with his characteristic grin, that he was "flexible, adaptable and agreeable." I was to learn the irony of that grin and to discover he was at best cautious, and at times the opposite of flexible, if not intractable. Yet that same grin charmed many a lady friend, including me.

Bill Dow, the first certified organic farmer in the state, was a profound catalyst to the organic farming movement in Chatham County and central North Carolina. He was wholeheartedly committed to his farm work, to the direct marketing of local food, and to his land. As you read these musings, imagine overhearing Bill's soft Southern accent, the cadence of his talking and his favorite expressions. Imagine his mischievous smirk and an occasional twinkle in his eye. You will discover the wisdom of a seasoned farmer, one for whom farming was far more than owning a piece of land. And you will learn a lot about how to farm, a way of life which he lived to the fullest and which he loved.

Daryl Farrington Walker (Ms.)
AYRSHIRE FARM, PITTSBORO

Foreword

Seven years ago, while working at Il Palio restaurant in Chapel Hill, I started to reinvent how I sourced my food and cooked. I owe that reinvention to my former executive chef, Adam Rose. He taught me the importance of using local, organic produce. To do that, he stressed building relationships with the incredible people that dedicate their lives to producing the food we eat.

Around this time, I met a farmer dropping off an order of amazing arugula. It was Bill Dow. He was wearing his infamous Bill Dow Boonie hat and was wet from a recent rain. As we unpacked the arugula, I asked him why there were so many small holes in the leaves. He gave me my first lesson on pest identification: "That's flea beetle damage. It happens to a lot of brassicas."

From that day on, every time I saw Bill, we talked about his produce, and agriculture in general. I realized I had never grown my own food. My wife, Whitney, and I began to reevaluate where and how we shopped and spent money. We built two 4' x 8' raised beds and grew eight tomato and twelve basil plants! Later that year, Whitney interned on Everlaughter Farm with Will Cramer and Sam Hummel, and I got a certification for sustainable agriculture at the Pittsboro community college. We started to farm in Hillsborough and sell produce at the Southern Village Farmers Market.

I spoke with Bill weekly about what was going on at our "farm" (a quarter acre vegetable garden on twenty total acres.) I could see how happy he was when he heard my determination and drive while discussing farming difficulties. He knew before I did that I would be doing this for the rest of my life. I mean, he is Bill Dow! He always promoted what I was doing and listened to everything I said.

One time at the Carrboro Farmers Market, Bill introduced me: "I want you to meet a friend of mine. I haven't seen a kid fall into agriculture like him in a long time!" He believed in me and what I was doing. Talking to Bill made me feel like it was possible—to be a full time chef and a full time farmer. He respected me as a chef and as a new farmer.

In the fall of 2012, my wife and I went to the CFSA Sustainable Agriculture Conference in South Carolina just before Bill passed. As we walked to the elevator to a workshop, we ran into Bill and Daryl. We smiled, gave each other hugs and made plans to sit together at dinner. I told my wife that I wanted to make sure we got pictures with Bill and Daryl, since "Bill's not getting any younger." I cherish that photo.

One month later, I learned that Bill had passed away in his home at the farm. It broke my heart. My first mentor was gone. I went to his service at the farm to pay my respects. It was a wonderful celebration of his life. We buried Bill beneath one of his apple trees. As Daryl started speaking about Bill, she placed his infamous Boonie hat in the hole. Bill's beloved dog, Katie, slowly walked up and laid down next to it looking in. That's when I lost it. I'm sure Katie knew what was going on. Daryl invited us all to take a handful of Bill's ashes and sprinkle them wherever we wanted on the farm. As I grabbed my handful, I felt something metal—a button from the shirt Bill was wearing. I kept that piece of metal.

A year later, I bought ten acres of land in Mebane for our permanent homestead. I decided to bury that small piece of Bill on my property, in the southeast corner, for the sun to rise every morning over Bill. I know that he sees what we have continued to do with our lives and looks over our farm with every sunrise. I love Bill Dow. He will never be forgotten.

Isaiah Allen
EXECUTIVE CHEF, THE EDDY PUB, SAXAPAHAW
OWNER, ROCKY RUN FARM, MEBANE

Editor's Note

Surrounded by hundreds of eager farmers and activists, I was sitting next to Bill at the opening dinner of an annual sustainable agriculture conference. The keynote address was by a West Coast author who gave a polished speech about his new book and organic farming projects. He talked openly about the struggles of his new farm and the beginner mistakes he had made. Afterwards, I leaned over to Bill and said, "Why is it that the books these days are always written by the newbies, like this guy or Barbara Kingsolver, chronicling all their newbie errors? We rarely see anything in print from the experienced folks like you." Bill replied, "It's because we experienced folks are way too busy to write anything down."

I went home that night and thought about Bill's comment. What he had said was true. He *was* too busy, and on top of that, he was not good with computers. He was more of a pencil and paper guy. One other thing I knew about Bill was that his health was declining. If Bill waited until he retired to start a writing project, would it be too late?

Bill and I became acquainted as fellow board members of the Carolina Farm Stewardship Association, the well-loved sustainable agriculture organization, which advocates for organic and small farms and which he helped start. He was the Board Chair and I, the Treasurer. From the start, I liked Bill. He had a sly sense of humor, a keen sense of mission and purpose, was very intelligent, confident, and even a bit mysterious. I could never quite figure out how he transitioned from being a medical doctor to becoming an organic farmer, how he landed in North Carolina from Mississippi, and how he became so passionate about organic farming. Considering Bill's role as a pioneer in the organic and direct market movement in the Carolinas and his influence nationally, knowing his story seemed important.

Over several years, I got to know Bill better, and my respect and affection for him grew. I was working in Pittsboro and it was easy to head down to his farm after work and hang out with him. At the time, Ayrshire Farm was for me a beautiful, peaceful place, with tidy raised beds, fruit orchards and cheerful workers picking in the fields. Some evenings I would help Bill and his crew harvest tomatoes or greens. Some summer nights, I would camp out at the farm to reduce my commute. He and his partner, Daryl Walker, would invite me for a delicious, healthy supper. I slowly came to know more of Bill's story, but mostly we just talked about the operation of the farm, the farm economy and politics.

With all this in mind, I decided that night after the CFSA dinner that I would make a proposal to Bill. I sat down at my computer and wrote out an outline for a book on which Bill and I could collaborate. I divided the book into four sections for the four seasons, and wrote a series of questions for each season, with dual emphasis on how the farm operated and the arc of Bill's career. The next day, I handed the outline to Bill. He read over it. "Let's get all that on paper and then you can edit at your leisure." Bill replied with a wink, "Well, I can hardly turn that down. We'll split the profits."

It took us two years to arrange the four interviews, each several hours long. The format kept us on track and gave us something of a deadline, although the seasonal approach was abandoned early on as too restrictive. After wrangling my way onto Bill's schedule, we would sit in cozy chairs in his small living room with one or two dogs wandering around. I had a digital recorder and analog back-up, and discovered that I knew very little about the technical side of oral history projects. Nonetheless, we ended up with about a hundred pages of solid text. (Anna MacDonald Dobbs did at least half of the transcribing.)

A paper copy of the text went to Bill about a year before his death. He was in the early stages of the review process when he died in December 2012. Subsequently, I have extensively edited the raw transcript, done some fact-checking and cleaned up the prose. Daryl has offered her edits and insights. I welcome any comments or corrections from our readers. I hope that the finished product would meet with Bill's approval.

As far as what Bill has to say, I hope that his humor, candor and passion come through. He was fun to interview. I had no idea he had such a complicated back story. I was happy to finally understand how Bill got so turned on to sustainable farming and turned off by the medical profession. I was able to let go of my erroneous notion of a rich family doctor turned hobby farmer. I applaud his vision so many years ago of a better food system and his hard work setting up projects and starting his own farm. As to his farming methods, he was idiosyncratic and not all will agree with his ideas (like questioning the usefulness of soil tests). I'm not sure I agree with his assertion that expanding direct marketing is the critical next step for a better food system as opposed to expanding wholesaling and larger farms. I see small-scale, direct market operations as the pioneers, greasing the wheels of change in the larger system. In that regard, Bill envisions a more radical shift than I do.

Regarding the title of his volume, I initially suggested "Shut Up and Dig." It was a reference to Bill's no nonsense side and penchant for action. I liked it, but Bill never did, feeling that it was too harsh. After his death, I came up with a new title: "What I Dig." This play on words reinforced for me the '60s counter culture origins of Bill's philosophy and captured his enthusiasm for the soil. I was happy with that one until Daryl pointed out that Bill never used that phrase and would not have liked it at all. The title that we settled on came to me while reading the moving tributes to Bill on Debbie Roos' website, Growing Small Farms. Lee and Larry Newlin had written this: "During open worship (at the Friends Meeting House) I had hoped to quote from Wendell Berry, but a lump in my throat from the sadness and tenderness of the worship silence prohibited that. What I wanted to share was Berry's, "I stand for what I stand on" and how that will resonate through the ages in the life of Bill Dow. What Bill stood for and stood on inspired countless wannabe and aspiring farmers, his customers at the farmers market and restaurants, and those who knew him in community." The quotation is from "A Part," a collection of Wendell Berry poems from 1980, and sums up what this volume is about.

I want to thank all the people who made this project possible, especially Bill's life partner, Daryl Walker. I thank her for her steadfast help, good cheer and encouragement, which got the project to completion, and for being a good friend. My appreciation also goes to Debbie Roos for her wonderful photographs of Bill. This is just one part of her amazing documentation of the Chatham County farming world. More photographs of Bill can be found at Debbie's website "Growing Small Farms." Thanks also to Anna MacDonald Dobbs for her excellent transcriptions; Tony Kleese, for introducing me to Bill and being a good friend and colleague; all the staff and Board at Carolina Farm Stewardship, especially Roland McReynolds and Cheryl Ripperton Rettie, who actively cheered on this work. A special thanks to Chris Crochetière and her colleagues at BW&A Books in Durham who did such a nice job designing the book. I also want to thank Biff Hollingsworth at the UNC Southern Historical Collection for assisting us to use the photos taken by Richard Davidson in the 1970s. Additional photographs can be viewed at the the website of the Appalachian Student Health Coalition Archive Project. In addition, thank you to the many local and organic farm advocates, farmers, chefs and food entrepreneurs who inspire me to work on these issues. Finally, a big thanks to my wife Debra Eidson and her son Ray, who offered steady encouragement during the long editing process.

And to Bill, we all miss you very much!

Fred Broadwell
DURHAM
OCTOBER 2015

Origins

From left to right, Bill's youngest brother, Staton; Bill; younger brother, John; and father, John.

(previous page) Bill's mother, Laura, and two younger brothers, John (left) and Staton (center).

Healthy Food

"At the heart of Slow Food's Triangle Chapter are dozens of farmers' markets, including the Carrboro Farmers Market, one of the most successful in the United States. The market was started thirty years ago by a doctor-turned-farmer who thought that providing good food was the best way to affect people's health."

That quotation is from an article by my friend, Moreton Neal, from the periodical *Garden and Gun*. Moreton is talking about me and her statement is not quite right. Good food is important to health. It is not the best way though. The best way, when all is said and done, is improving people's income. That's the bottom line. But ideas have their own particular time and this is the time for direct marketing of good food.

Let me take a step back. An example I like has to do with breast feeding. When I was a medical intern and resident, we had all the data we've got now about the benefits of breastfeeding. But I could have stood on my head and spit wooden nickels, and the response would have been flat out "I ain't gonna do it." "I don't want to and grandma says it's bad." There were a thousand reasons not to, and it wasn't going to happen.

But now we jump twenty, thirty years into the future, and everybody is breastfeeding. We don't have any more data than we had before, but it became the thing to do. There are certain practices that society takes on—not for necessarily good or bad reasons; it's just that it's the time. For breastfeeding, it's that time.

The time has also come for healthy food. The data has been clear for about twenty or thirty years. It had become fairly obvious to people that something was wrong with such prevalent obesity, hypertension and other diseases. Everybody was eating two or three meals everyday out of the home. Somebody pointed out that maybe what they were eating in those two or three meals out wasn't so good for them. People asked: what's the alternative? The alternative is to go back to the way we were before.

A lot of folks are saying that it *can* be done. There can be a relationship between farmers and consumers, and that very relationship can be good for your health. What is the quality, variety and freshness of the food we eat? How is that food grown? And I'll just say it. I do not believe that there is sufficient data to verify the human health safety of pesticides on food crops.

Bribes

When I was in medical school at Vanderbilt, you might say that I burned the backsides of some of the folks in the administration and faculty. I was unhappy, because there wasn't much talk about real people, their situations and the social side of things. It was very academic. To be a good doctor, it's not just a matter of blood counts and your level of cholesterol.

It was 1968. The medical school dean, John Chapman, came around to see me. He said, "There is an organization in New York called the Josiah Mason Foundation. It is going to put on a meeting to look at whether the radical medical students in America are going to take over the medical schools. Would you be interested in going?" I had never been to New York City and they were going to give me a ticket to go. Hmm. I didn't think I was a radical medical student, but if I could go to New York, I said to myself, "Let's go."

There were fifteen of us—each from a different medical school around the country. Dean Chapman was right about what they wanted to know: were the radical medical students going to take over? Of course, it was a ludicrous question, because if you were a radical student, you wouldn't have gotten in.

If you became one when you were in there, they'd kick you out. So they were naïve.

But to make a long story short, they were indeed concerned about medical students causing trouble. For a payoff of a grant of twenty or thirty thousand charitable dollars to keep us off the streets, we would let the foundation "keep us out of trouble." I didn't mind being bribed. So before I left New York, I had hatched up an idea. We would use the money to find out about health conditions in rural areas, in my case rural Tennessee.

To figure out how and where to do a useful project, we first had to do some homework. The underserved population in east Tennessee was primarily white and lived back in the mountains. The underserved population in west Tennessee was primarily agricultural and black. We had one of the two black medical schools in the country in Nashville; it was called Meharry. It was suggested that we work out some joint activities with them. We could do a survey to learn what was needed and what might be done. I said, "Fine." When I got back home, I started to make the project happen, rather than just talking about it. It took off.

Fluoride

About this time, I had the good fortune of finding Dr. Amos Christie who had been the head of the Vanderbilt Medical School Pediatric Department for years. He had just retired and had time on his hands. This fellow didn't mind being a burr in folks' saddles. He was a good guy. He was the one, at least by his rendition of it, who integrated the hospital there at Vanderbilt. It used to be that the wards for black infants were in the basement. He came in to work one morning and went down to the basement. He picked one up of those babies and walked upstairs, put it in the crib and that was that.

Dr. Christie was willing to stick his neck out. I can remember him saying, "Boys, what you do is your business. I just need to know what it is, so that when I go to the bank, I know what

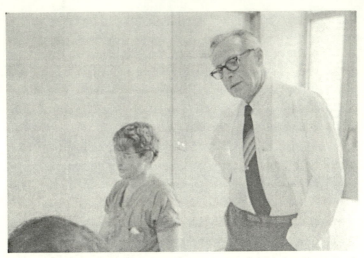

Dr. Amos Christie (right), and Bill, at Vanderbilt University Medical School in 1970. (Photo by Richard Davidson, courtesy of Southern Historical Collection, UNC-Chapel Hill.)

to say." He wasn't literally going to the bank, but people were asking questions. I was fortunate to find him and so were other medical students.

For the first month and a half of our project, we headed down to Lincoln County, just south of Nashville, and examined Head Start children. It was a rural county, but exposed to all the stuff going on in Nashville. Our group consisted of first and second year medical students, nursing students and two recent nursing school graduates. Dr. Christie taught all of us how to do histories and physicals. He made no distinction between the medical students and the nursing students, which made an impression on me.

Next we went out and examined Head Start kids in rural Williamson County, and a couple of things became apparent. At least half the kids had parasites. The old adage was that kids didn't have parasites anymore. The Rockefeller Foundation had taken care of that. Surprise, surprise! Probably it wasn't as bad as it had been, but the parasite problem certainly wasn't solved. The other thing was the children's dental health. By looking in their mouths, you could tell whether the children were from out in the country or from in town. If they were from in town and had been drinking fluoridated water, they didn't have nearly the dental caries as the kids from out in the country. It was just glaring. It wasn't just a couple cavities here and there, it was terrible. Dental caries has serious ramifications; it's a latent infection that sets you up for a lot of other illnesses. It's a real problem.

What was going to happen about all that? The worms weren't that big a deal, because the health department was willing to recognize the problem. We had well-documented findings. That issue got taken care of. But the dental caries was another matter. Obviously, there wasn't going to be a water system out in the rural areas that could provide fluoridated water. But you could do a topical application of fluoride. It's just like a toothpaste—you put it on a couple times a year and it cuts dental cavities by a third.

We went to the dental supply place and got a couple of tubes of fluoridated toothpaste. We got some toothbrushes from a toothbrush company and proceeded to visit the kids in their homes. We would do it with the whole gang, not just the ones we had previously seen; if there were other kids around, we'd invite them, too. We would show them how to brush their teeth and put on this topical application of fluoride.

One day we got a call that the sheriff was on his way to get us. We had better get out of Williamson County. "What do you mean the sheriff is after us?" Well, you're practicing medicine without a license. Or practicing dentistry. "What are you talking about?" Nonetheless, we boogied and got across the county line.

After we got to looking into it, this is the odd truth we learned. The head of the local health department was the retired, longtime head of the State Health Department; his son was the current head of the State Health Department. In the past, the father had studied this particular situation personally and knew the results. He knew that topical applications of

Bill during the student health coalition days, rural Tennessee, 1970. (Photo by Richard Davidson, courtesy of Southern Historical Collection, UNC-Chapel Hill.)

fluoride decreased caries by a third. He knew, but had taken no action. He'd put the study on the shelf. It was this father and son who called the sheriff.

I'm a little slow about things sometimes, but at that point I said, "Hey, wait a minute, what the hell is going on here?" If I hadn't been somewhat irritated with the system up to that point, well, that did it. That was blatant, inexcusable, or whatever you want to call it. I think it lit the fire under me.[1]

---

1. I recognize that the medical profession is a guild with rules, and if you are going to be in it, you have to follow the rules. I don't know how much I recognized it up to that point; I certainly did then. If you want to set me up between rules and the kid, the kid gets it every time. You may have to change the rules, or you may have to do some bargaining, but you don't bargain the kid, you shouldn't. But it does happen all the time. Look at the present health care situation.

Where the Coal Was

Getting run out of town by the sheriff pretty much finished us off as far as the Head Start kids in Williamson County were concerned. There were a couple of articles in the newspaper. The local health department got off its do-nothing stool, but it took a while. There were some frayed feelings.

I spent the last part of that summer in east Tennessee with one other medical student and a nurse, Brenda Oakley, whom we met through a community worker named Reese Roe.[2] Brenda was putting on what folks were calling "health fairs." These were sponsored primarily by the Presbyterian Church. She had a station wagon filled with posters, a card table and supplies, and was traveling around old coal camps in Tennessee and Kentucky.[3] I had the utmost respect for Brenda's work and it proved very helpful to us.

---

2. Also John Aventa, Rhodes Scholar and teacher, came back to Vanderbilt to help about this time.

3. I remember being in Hogg County one night. We were supposed to meet Brenda the next day at a place called Everettes. We got into town and it was close to midnight. (It was in the town of Harlan.) Problem was we didn't know how to get to Everettes. The only light on was in the Sheriff's Office, so we pulled up. The guy with me was Pat Maxwell, a nice guy and a big guy, but not real experienced in the ways of the world. We walked in and this deputy was sitting on the far side of the room behind a desk, with his feet up on the desk and his revolver in his hand. He was cleaning the revolver. I guess Pat liked guns. He walked over, and after we introduced ourselves, he just took the thing right out of the guy's hands. "Nice gun," he said. The deputy sheriff's mouth dropped open and my mouth dropped open. We managed to talk our way out of that one. In the end, the deputy himself took us up to the place we were supposed to be staying in Everettes.

Driving up there, it's a mountain road, and there is the down side and the up side. The house where we were staying was on the up side, but we did notice a little house on the downside, because we were looking for houses. We got to the right place and they took us in and put us in a bed. Next morning, we got up and walked out to the front porch, and the little house down below was gone. I could have sworn there was a house there, I said. I looked around and there were the remains smoldering. I asked what happened. They said old so-and-so had a fire last night, and the place burned down. What do you mean? It turned out that he was bootlegging and had passed off some bad stuff. This was a different way of handling things.

Growing up, my family had a farm in Kemper County, Mississippi. We were way outside of Brady, the second largest town in the state, at 50,000. Boot-

Brenda's health fairs were drawing some folks, but not exactly crowds. By our estimation, the fairs needed two things. The medical part needed to be booted up to do a lot more than simply talk to kids about a few health topics, which was all that she could do. The fairs also needed local sponsorship. She had done some of that work, but not nearly on the scale that it took to make things rock and roll. It was clear that it could be a good project if we revved it up. The result of our revving became what was called the Appalachian Student Health Coalition, which did in fact put on health fairs with a lot of local sponsorship.[4]

We hired a couple of students to look at social programs like Medicaid and Medicare, and to help folks learn that these programs were available. A number of communities eventually set up their own health services, built their own clinics and hired staff. But, whether it was out in western Tennessee or up in the mountains, the quality of health outcomes had to do with funding. The existing medical facilities needed money. The health fair could find that Uncle Fred had high blood pressure, but if Uncle Fred didn't have the money to go get something for it, we weren't going anywhere.

The funding problem took a turn in 1971. One of my early heroes was Nat Caldwell, a Pulitzer Prize-winning reporter for the Nashville *Tennessean*. This is the guy who put the finger on John L. Lewis for being on both sides of the coal issue—Lewis had been both an owner and a union leader. Nat knew so much about the TVA and the South, and he was just a great old guy. He always wore a suit and tie, yet still looked quite disheveled. By God, he was a bulldog. He'd get into stuff.

In 1971 I was out at Nat's one night, and we were sitting around talking. "Nat, where are these communities, these old

___

legging was a major issue there, too. So I wasn't all that ignorant about that kind of thing. Fire was a weapon: your pasture got burned, and the next time, if you didn't learn your lesson, something else got burned. In Everettes, at any rate, everybody was kind of blasé about it.

4. Today the coalition lives on as the Center for Community Health Solutions, based at Vanderbilt.

coal communities, going to find the money to put up health facilities?" He said, "I'll tell you where the money is. What is going on up there is that the land that's got coal under it is not being assessed for that coal." State law said that you had to assess what's on the surface plus what's underneath. What you would find in the records, Nat told me, was that the tax office was only charging for what was on the surface. Those counties, and there were five major coal-mining counties in Tennessee, were losing a lot of money. "Can we organize around that?" He said, "That's your problem. You be the organizer and I'll just tell you where the money is."

I reported this back to our team, and we decided to investigate. Our three interns (John Deviner, Alan Allman, and Bob Thompson) plus some other helpers spent the summer of 1971 up in those courthouses in east Tennessee. They went through the tax books and found out what Coal Company X was paying in taxes for land that was adjacent to Farmer Y. They discovered that the farmers and the coal companies were paying the same per acre. When our team asked the counties why this was, we were told: "We don't know where that coal is." Now wait a minute! Obviously somebody bought this piece of land because they knew where the coal was. There are coal maps. "Aw, no, no, no," was the response.

They did know where the coal was. The potential revenue from those unpaid taxes could do a whole lot for those counties that were starving for funds. It could fix the bad roads, bad schools, bad social services and the lack of medical care. The wealth was there. It was being trucked out, though, with no tax revenue in the bargain.

Two major in-state owners of coal lands were Senator Howard Baker and Senator Albert Gore, Sr. Howard Baker had his heart set on running for President. Gore later became head of Iron Creek Coal Company and had a famous son. All of a sudden, this little project we were dinking around with is making a splash. It was like us being down there in Williamson County. We were just dealing with little kids with toothbrushes and what the hell is the problem? Now we're just dinking around trying to get some folks some health care. What's the problem?

When the summer was over and we had the data back about the under-payment of taxes, we went to John Sigenthaler, the editor at the Nashville *Tennessean*. He had a lawyer friend who took the case. Members of a citizens group from east Tennessee, Save Our Cumberland Mountains (SOCM), at great risk to themselves, filed suit against the State for not enforcing the tax laws, eventually winning the suit. J. W. Bradley was the head of SOCM and not one to dally around.[5]

Thinking about today's situation, I can't help but say that mountaintop removal is a travesty beyond the imagination. To go up there and see what is being done today is shocking. The coal companies are allowed to literally take the top of the mountain off and then down the sides. Such damage to your home place does something to your health and to your mental health. There are sludge dams breaking, and people being killed. What happens to you when the only place you can get a job is in the mines and that is connected to such destruction? What happens to you when the vista you've been able to see all your life is gone?

Rural Breakdown

There is no question that money was the central issue. If you looked in rural areas in Tennessee or wherever in those days, the one thing that rural folks had going for them was they knew how to farm. Now, to be more accurate, they may not have been farming much for the last generation. They may have come to believe that it could not be successful. But they still had land,

---

5. The success of this lawsuit helped launch Save Our Cumberland Mountains, a well-respected coal country advocacy group. The organization later changed its name to Statewide Organizing for Community Empowerment (still abbreviated as SOCM) and is today an active statewide community group with multiple staff and local chapters. A short video on the history of SOCM is on YouTube.

some of them, and they still had some of the equipment, and some still knew how to do it.

So where was the breakdown in the system? Why was it that these folks didn't want to grow anything and as a result remained in poverty? Here is the answer. It was because they had been growing crops that now were only profitable at a larger scale. Unless you grew a lot, you couldn't turn a profit. That had been the practice at my home, growing up. We were growing cows and soybeans on 150 acres. You can't compete with somebody with 5,000 acres; they can undercut the price. The big guys can always undercut the price and leave you out. If you have to sell at the big man's price, you're not going to make it. Also, as many people are now aware, between the farmer and the person doing the eating, there are a lot of folks with their hands in the till.

Back then, I wondered: if you could cut out all the people in between, farmers ought to make some money, at least on paper. And if the people in rural areas can start making more money, then better health care is possible, and hopefully diets become better, too. It would be hard to defeat Ho Hos and Twinkies, but it was certainly not impossible, I thought.

After medical school, I became the director of the Center for Health Services at Vanderbilt and we were fishing for an agricultural marketing project. About that time, the National Farmers Organization (NFO) came along. One of the things they did was sell ground meat off of trucks. They set up in parking lots in the suburbs. It seemed easy and it seemed to work. I thought, now there is an idea. Meat isn't all that folks got, but this really works.

In those days, vintage farmers' markets still existed, but they were located in downtowns. The one in Nashville was downtown, for example. The problem was that shoppers didn't go downtown anymore. They had all moved to the suburbs and the supermarkets with them. Kroger, Food Lion, everybody, they had all moved out. The farmers who had stayed downtown weren't making any money; they were selling wholesale now, instead of retail. My thought was to get the farmers where the customers were, to sell retail again and make money. Put farmers out there next to Kroger, undercut their prices and we can whip Kroger. The economics weren't hard to see, but not everyone agreed.

When we hatched up the idea to put together farmers' markets, we had been doing work with the Student Health Coalition for a while and people knew that work. But starting farmers' markets was a new idea. Who would support a project like that?

We soon discovered that the Field Foundation was one of those groups that helped people initiate things. The foundation was led by Les Dunbar, a fellow who had lived in Durham for a long time, and written about Appalachia and Southern history. Les might be the right guy for us, I thought. I called up Les. He said, "Well, I'm going to be down that way. Why don't we meet somewhere?" We met halfway between Nashville and the mountains. I remember it was a round table with a lazy Susan in the middle of it. We told him what we wanted to do with these farmers' markets. He looked at me and said, "Do you really think it's going to work?" I said, "Yeah, it's going to work." "Alright, let's do it." That was it! I laugh about it to this day.

Now, granted, the idea was pretty good. But the other thing was that we had done what we said we would do the year before. So with that knowledge and our brief conversation, the Field Foundation helped us get started and then go on to get money from other places, other big foundations. I can say that there is a straight line from what Les said that day to the Carrboro Farmers' Market out there—a very straight line. People like Les need to get credit for sticking their necks out before all this was popular.

Dirty Old Farmers

This story tells you how times have changed. When I was in Vanderbilt Medical School, I had gotten to know the woman who was the secretary to the Dean. She had a daughter in high school who wanted to be a forest ranger. That was a pretty dodgy post at that time, especially if you were female, so I was impressed.

The first time we ever had a farmers' market there in Nashville, my friend and her daughter decided to go to check it out. I was real interested to see what their reaction would be. They represented fairly typical Tennessee folks, bright and a couple of generations off the farm.

The following week I went to see her and asked, "What did you think?" She said, "Well, it was great, but my daughter was real upset about it." "Well, what happened?" "When we got down to the end of the line, she took me aside and said, 'Momma, you're not going to buy food from those dirty old farmers, are you?' "

Oh, my God. Little things like that, some of which you remember and some of which you don't, are what boot you into realizing where you need to be going. You're not reading this out of a textbook or following somebody's formula or strategy. It's a matter of listening to what people are saying and realizing, whoops, we gotta do something about that.

My friend's daughter didn't have an issue with the food; it was the farmers. They had on bib overalls and brogans and probably had been in cow shit before they got in the truck. Those aren't the people who are farming anymore. Most are dead and gone—not all of them, but a lot of them. Today, the dirty old farmers probably don't look quite as dirty as they used to. We owe a lot to the dirty ones who came before us.

Treflan

My shift to direct marketing and organic production had actually begun through arguments with my Dad, with what had been going on at home. Neighbors would call him up: "Mr. Dow, would you like to sell some meat to me?" "No, no, we don't do that. We take it to the stockyard and sell it there." He just flat out would not do it. Now I could sit down with a pen and paper, work out the math, and say, "Look." But the answer was always, "Nope, nope." That is what he knew. That is what a fellow ought to do.

Being the oldest son, I have had some differences with him along the way. Given the climate of the '60s and '70s, there were a lot of differences. So I have to admit that while mostly I was motivated by wanting to find ways to get money back in the rural areas, part of the attraction of this direct marketing project was some of that old spiteful, "I'll show you."

The price of gas was going up and so was the price of fertilizer. I thought to myself, "Do you really need that fertilizer? Do you really need all that stuff?"

I'd gone home when I was a college student to help plant soybeans. We were using a pre-emergent herbicide, Treflan. After I was in medical school, I would still help at the farm occasionally. At this point I said, "Boys, I will help out, but I'm not spraying the Treflan. You can do what you think is necessary, but I'm not doing it." And that was a major break in the family. Here I am in medical school and I'm not willing to do this. It really threatened the way that they saw things. Both my younger brothers went to ag school. I didn't, obviously.

So why didn't I want to spray the Treflan? Well I ask you: is it good for you? Look at the damn label. The manufacturer talks about all these things you've got to be careful about. And sure the data said that it hasn't caused trouble, but what I want to know is will it cause trouble in the future—is there proof? And that is not the way that the data was set up. In fact, when they began to look at the studies, two thirds of the chemicals on the market didn't have enough research findings for what is called a Health Hazard Assessment, the minimal research needed to let a product on the market. Two-thirds!

Treflan was a pre-emergent herbicide. It was to keep the weeds out. What else could we have done? As far as the chemical companies and my family were concerned, there wasn't any alternative. "That is the way you got to do it." But I'm thinking there's got to be other ways of dealing with the weeds. I don't know what those are, but it's about time to find out.

Here's one of the things that we did find. The U.S. Geological Survey was looking at an old tobacco field in Guilford County. They were looking at both surface water and ground water, checking on whether there were chemicals in there and if so, what they were. What they found, both in the surface water and the ground water, were chemicals that have not been used in the last twenty years—they were still in the ground! That's scary. Real scary. So even if we change what is being used today, the old stuff is still there. Shouldn't we find out more? Suppose the initial studies weren't really very good or were corrupted and somebody fudged the data. We put all that chemical stuff out there and Lord knows how long it could still be causing problems. Twenty, fifty, one hundred years later? Is that not a reason to be cautious and perhaps to grow organically?

If I put a salt shaker on the table and tell you that Treflan is in there, and let's shake a little on your soybeans before you eat them, you would say, "Wait a minute!" But by serving soybeans to you with no sign on them that says they have been treated with Treflan, you don't object. You'll walk out and feel just fine. I don't expect you to keel over. But at some point, there may be a problem with that. I don't think people want to knowingly take that risk.

That is what we are seeing now with the food movement. People are changing how they are eating and who they are buying from. They are buying food from people whose farms they can go out and see, farmers they know personally. On the farmer's side, I don't feed something covered in chemicals to someone who is my friend.

There is a whole new ethic that is evolving about agriculture. It has taken a lot longer than I thought it would, and we haven't converted everybody, that's for sure. But it ain't what it used to be! There are markets up all over the place that are rockin' and rollin'. The problem now is production: there isn't enough production, which makes it a very different situation than where we were ten years ago, fifteen years ago.

Mississippi

But, of course, I am not an organizer myself. I'm a farmer. This goes back to my growing up on a farm. Now most of us are genetically equipped to do farming. It is the reason we have survived. If we couldn't do it, we wouldn't be here today. So genetically, we're there. But some of us, thankfully, just like doing it. We might cuss at it and angst over it and everything else, but we still love it.

When I left Mississippi and the family farm, I did miss aspects of it. There were things about it that were good. There were those times of doing things together. There aren't a lot of things that families can do together and farming is one of them. One of the few. That doesn't mean we all got along. Hardly the case.

Probably the most influential thing in my life growing up was that my folks bought a hundred acres of old worn-out cotton land. It was up in Kemper County. And I mean worn-out. They didn't pay much for it, because it wasn't worth much. It was grown up in pine and sweet gum. This was the early to mid '60s.

My father went to town and bought some axes and cross cut saws, and said, "Boys, go to it." My mother was scared to death of power equipment, and we were not going to have a chainsaw. You can beat yourself up about that, but in the end it makes for a better story. I am sure there were many times I thought I would rather be doing anything else. It was hard work. But in fact, and I've talked to my brothers about this, in retrospect, clearing that land was the most important thing that happened to us when we were kids.

When my youngest brother came along, we produced hay on the farm. My brother, John, was on the state championship football team, and they were good. John managed to hoodoo most of the guys on the team into coming up and working on the farm, putting up hay and helping out. Any time you get around any of them, even today, the first thing they start talking about is the farm. It wasn't theirs, but you would think from the way they talked that it belonged to them. A farm is more than just a piece of land. You hope you make a living out of it, but there is something else there, too.

My mother was a great gardener. Vegetables were not even a challenge for her. She always balked at using chemicals to garden. That is not to say that it never happened, but she was on the side of "let's be careful here." One other thing she worked on was getting native wildflowers to grow in the garden in tough environments. That kind of interest made me think that somewhere in her mind she was concerned about the agricultural chemicals on the farm. There was too much invested in what was going on for her to make any real commotion. Nonetheless, I think she was worried.

Eliot Coleman talks about inputs for the farm.[6] You can create them on the farm, or you can get stuck buying them from the salesman. The salesman is very persuasive. He rules agriculture in many respects. Chemical fertilizers, chemical herbicides, and insecticides. Just look at the names. *Attack. Bravo.* It's not hard to figure out why they choose those names. I think of Jerry Clower, the comedian from Mississippi who lived over in Yazoo City. He had been a chemical dealer before becoming a famous Grand Ole Opry comic. The jokes and the humor that went through all that selling.[7]

Medicine's got them, peddling pharmaceuticals and medical technology—they're called the detail people. Agriculture's got

6. See *The New Organic Grower* and *Four-Season Harvest*, two of his classic books.
7. From Wikipedia: Howard Gerald "Jerry" Clower (September 28, 1926–August 24, 1998) was a popular country comedian best known for his stories of the rural South. He was often nicknamed "The Mouth of Mississippi." After finishing school, in 1951, Clower worked as a county agent and later as a seed salesman. He became a fertilizer salesman for Mississippi Chemical in 1954. By this time, he had developed a reputation for telling funny stories to boost his sales.

them, peddling patented seeds and Roundup—they're called the chemical people. They're running all of us. I could be very expressive about that, but will refrain. They run the show. They're slick. They've got big money behind them and they show that their system works. Nobody wants to dig into it too far, because maybe it isn't what we think it is. Those soybeans look good and no weeds in them either. But do I really want to eat them? Now, hmm, that is something different.

North Carolina

The Vanderbilt-based health project set up markets in Tennessee and that was good. Then, in 1978 I came over to North Carolina on a fellowship with the Johnson Clinical Scholars Program at UNC. We set up the Agricultural Market Project (AMP) and the Student Health Coalition (SHC) in that same year. Frankly speaking, the fellowship was boring, oh my goodness. It was supposed to look at how to change health care systems, but there was too much pushing me on the agriculture front to be willing to sit there and listen for very long. I wanted to get moving.

At the end of that year, our colleagues in Alabama asked me to come down and work with them to put together an umbrella piece for the various Southern health projects. Part of the reason I agreed was that my father had been suffering from heart disease. He had had several major heart attacks and was in congestive heart failure. He was still getting around, but was not in good health. As far as our relationship went, we were winding things down, it was clear. He and I hadn't always gotten along very well, and I decided to straighten it all out. I needed to be down there and not somewhere else.

I spent a lot of time over in Mississippi with him and working with folks at the University of Alabama-Tuscaloosa. Dad died in March as I recall. It took a long time to go through the boxes and boxes. I was planning to leave, so I felt the time pressure, unlike my brothers who were sticking around.

I had been seeing a woman, Leah, whom I had dated in times past. She had become a midwife, was a nurse practitioner, and was on the faculty of a program in Rochester. There was a very fine midwifery program up there, but she was ready to leave that town, and I was not planning on staying in Mississippi. So I said, "If you can find a job as a midwife, I'll go anywhere you want." I could take my training with me wherever. Looking back, this was a mistake, at least in some respects. I told her that there was a birthing center in Siler City and that she ought to get in touch with those folks. Once the cat was out of the bag, it was a done deal. I didn't want to go to North Carolina. I wanted to go back to Tennessee.

The Tennessee projects remained up in the air and the folks in Siler City offered Leah a job. Next thing I know, we are talking to Elizabeth Anderson, a Chatham County real estate agent. She knew about this place for rent. We came to look at it and boom! For a couple of years, we rented the farm. After a while, Leah and I realized that we just couldn't live in the same house together. She is a very important person to me, but it just wasn't to be. So I stayed and she left. She is in Albuquerque now with a husband she met here, who has New Mexico roots. If you pick up a midwife journal, you will find her name in there somewhere. She's made a tremendous contribution. Long story short, I bought this place.

Down South and Up North

When I first bought the farm, it had grown up a bit. It was not as bad as the situation in Mississippi, but there were definitely lots of sweet gum and pine. I spent some time really cleaning it up. The place where the number four terrace is now, right in front of the shed, was the flattest place out there. I tilled that, planted a garden and it did well. Oh boy!

My early dealings with Cooperative Extension in Chatham County were, to put it mildly, not so good. Our current Extension agent, Debbie Roos, who is very good, will dislike me saying this, but it is just the truth. Down home in Mississippi, we had an Extension agent, Bobby Ross, who was a chemical guy. In those days, everybody was chemical. But there was a good relationship between him and the folks. He was somebody the local farmers really looked to for help. Based on this experience, I had a good impression of Extension.

When I went to visit the Extension office here, they sent me around to see an agent. It was about the time I was catching on to the idea of organic agriculture. This was partly because of what I had seen going on with farmers at the farmers' markets and my uncertainty about the chemicals. So I popped in to see the Chatham County agent, thinking that Bobby Rosses were everywhere and we could sit down and talk. I told this guy what I wanted to do.

He said, "Well, you down South now, boy, and I don't think that kind of stuff will work down here." I said, "No, no, no, you misunderstand. I'm up north. I'm from down in Mississippi and I'm up north, and I think we can do that up here." That is about as far as we ever went.

Big-Headed Broccoli

In my second year farming the land, I started growing broccoli, which was something I had been growing in my garden back in Nashville. One day this question came to me: can I get the supermarkets and other larger retailers to buy from me? So I tried my own little experiment. I offered most of my broccoli crop to a produce guy in a supermarket, and, lo and behold, he said yes.

I had grown central head broccoli with a large head on it, and it was nice looking stuff. Premium Crop was the variety. The supermarket guy would buy everything I had and at a good price. He was helping me, and I was helping him. It was good to learn something about the retail side of things.

I came in there a few days later, and he laughed and said, "I had a complaint about your broccoli this week."

I said, "What are you talking about?" "A woman came in here and said that you been cheating me on this broccoli."

"How do you figure that?" "The woman said, 'Well look here, on this one there is only one stem. I usually get four stems!' I just laughed and said, 'Oh, my God, ma'am! Are you buying it for the stem or the head?'" Here was someone who hadn't a clue as to where the value of that broccoli was. That is not to say that the stem doesn't have some value, but most folks are not after the stem. Here is a farmer providing glorious broccoli and the buyer doesn't have a clue. The education on the consumer side can't be left out.

Carrboro

Not too long ago the Carrboro Farmers' Market had its 30th anniversary. Let's talk about the early days, as best as I can recall them. Initially, two of us were especially interested in starting a market: me and a student named Alan Baumgartner. We pulled in folks from the UNC Medical School and the School of Public Health. Bobbie Wallace was hired to be the organizer for Chapel Hill. The first market was held in the parking lot of the Church of Reconciliation, which is near Eastgate Mall and just north of Franklin Street in Chapel Hill. The good thing about working with the church was that it provided a set of consumers with some commitment to the market. At the time, the Church of Reconciliation was not a big church, but it worked nonetheless.

At first, we felt very lucky to get a dozen vendors. There was a lot of reluctance. We practically had to go get them and carry them up there. (One of those people died the week of the anniversary: Henry Sparrow, a great guy who sold next to me at the Carrboro market for years.)

I can mention the names of a few of the original vendors. Wilber Bryant, who sells here in Pittsboro, is African-American and quite a personality. He has been selling for a long time. I recall Sarah Llewellyn and Jack Hanton. Other names escape me. Everybody deserves to be listed in an official history of the market.

The market got bigger the next year, so we moved to the Savings and Loan lot that is adjacent to the church. The market grew more, so we moved over to Eastgate Mall. By that time, there were also a lot of requests for help from other communities to start markets. Richard Pitman and Dale Everest did a lot of organizing in other places.

We brought the first set of rules for the market from Tennessee. The market was created for farms located within fifty miles of Carrboro. Rule number one was if you don't grow it, you don't sell it. All the rest of the rules were somewhere way behind that one. There could be no exception to that. In fact, one of the original people who became a market manager was kicked out for that very reason. Another one of the original vendors was kicked out because he never could quite distinguish what was his and what he got from somebody else.

And here's why that rule was so important. Take my friend Henry Sparrow, as a good example. Suppose Henry's got five or ten acres of corn out there. He's got money in it, he's taking care of it, and it's ready to go. He takes it up there to the market. The same week one of the other farmers realizes that it's time for corn to come in. He goes down to Georgia or Florida with a big truck, buys up a load, comes back up here, parks his truck and sells. He can cut the price, because he got it cheap. Henry's got some corn, good corn, but his price is not that low, because it cost him more to produce it. A couple times of that happening and it would just kill the market.

I recall that there was a guy who had been selling pecans at the market for some time. The freeze last year got everybody's pecans, including those in his neighborhood. But he kept bringing pecans. Needless to say, there was a problem. The committee that oversees that issue asked the vendor, "Where did the pecans come from?" Whether true or not, the committee concluded that they weren't his pecans.

Another example involves a guy who sold a lot of landscaping plants—this was back in the early days. He would go down to Georgia or Florida, pick up some plants, bring them back up here and eventually bring them to market. The question was how long should he hold on to them. How long did he have to hold that plant before it became his? The board didn't like what was going on, he disagreed, and it was a very contentious issue. No easy answers to that type of situation.

An important issue was making sure that the market didn't become another craft market. So we had to restrict the number of people with crafts. We've only got so much room to begin with. That's not to say that there's anything wrong with crafts; it's just a different kettle of fish. If somebody wanted to sell a few of this, that or the other thing along with their produce, that was OK. If somebody wanted to come and make it a flea market or a craft market, that was different.

The next phase of the market was the move from Eastgate Mall to Carrboro and then the building of the market structure. Mike Brough was Carrboro's town attorney at the time, and he knew about a source of federal funding to build market buildings. The Town of Carrboro wasn't overly receptive to the idea, but Mike pulled a group of us together anyway and we got to work. The big question was where we would put the thing.

Carr Mill Mall had just been developed, and there weren't enough parking spaces within that area to fulfill the town's requirement. The developer owned another lot behind Carrboro Town Hall, which was flat and could accommodate some extra parking. A deal was struck, and the land behind town hall was donated to the town, creating more parking plus a market space. In this way, the federal government paid for a roof over our heads. It all happened rather fast, with Bobbie Wallace doing a lot of the legwork. And thus, the Carrboro Farmers' Market got its permanent home, next to Town Hall.

The agreement was that the Town would be responsible for the shed, while the farmers would have complete control over the running of the market. We would pay rent to Carrboro. The market did well in its new location and the shelter was wonderful to have. The main drawback to the location has turned out to be that we have been handicapped with limited parking. It is the big issue, especially for older people. We just need more parking.

Ayrshire

The farm is called Ayrshire. Why? The answer to that involves genealogy and a mistake. Genealogy is an addiction (for me). It's your own mystery story. People read mystery stories, but you've got your own. That's a reason to get out there and explore.

I only had one grandparent growing up and that was my grandmother on my Dad's side. She was not a great storyteller, but she had a lot of information if you sat with her long enough. She lived in Madison, Indiana, which was a little town on the Ohio River that stopped growing about 1880. As a little kid, I spent time in the summer with her and it was a Samuel Clemens idyll. The Delta Queen stopped there. They had the steepest non-cogged railroad in the country. There was a guy who had a shop on the river where he built Chris Craft pleasure boats and also a hobby shop (because the boat yard would burn down occasionally) where I loved to hang out. The Dows and the Watlingtons (my father's maternal side) all came to Indiana when that part of the country was just being opened up right after the war of 1812—the Dows from Scotland and the Watlingtons from England, Bermuda and Scotland.

As a young man, I found a letter that my grandfather had written to his sister saying that he thought the family had come from Ayrshire, on the west coast of Scotland. That's where the poet Robert Burns is from, which is probably why he said it. In the early 1980s, I headed over to Scotland to have a look. When I got to Scotland, I went west and looked in the records in Ayrshire County parish. I was disappointed to discover that there had never been a single Dow in the County! It's not that there weren't very many; there weren't *any*.

So, then, where did these people come from? It turns out that my ancestors were actually from Perthshire, which is over in the eastern side of the country, north of Edinburgh on the Tay River. On the coastal plain as you move away from the water, right up to the Highlands, there is a fifteen-mile stretch of land and a parish called Moneydie. It is a breathtakingly beautiful place. In that parish, I found them in the records back to the 1740s: namely, William Dow of Northleys. "Where the devil was Northleys," I wondered. I looked and looked, and couldn't find it anywhere.

A few years later, I went back to Scotland. It was my wintertime farming break and I was there by myself. I headed to a little community called Octagarven or Bankfoot. It was a Sunday, snowy and cold; the wind was blowing hard. I saw a little Presbyterian church and thought: Somebody over there is going to know where Northleys is. So I returned to the church at eleven o'clock in the morning, in time for service. As it turned out, I was late. They had started at 10:30 and as I walked in, they were right in the middle of things. Everyone turned around and I thought, "Whoops, I am in the right place. They all look like Dows to me!" I had to laugh. When the service was over, I introduced myself to the preacher and told him what I was looking for. "Well," he said, "there are still a lot of Dows around here. And Northleys is one of the local farms." You were known by the farm that you were on—in other words, feudalism. Even though the agricultural revolution had begun, my Dow ancestors were still caught up in feudal traditions.

I got hold of a map and drove over to the general area of the farm. I took that map and just started walking. The map showed the boundaries between the farms; so I got over there and walked right up to the place. It was a time that I will always remember. Strangely, there is a large standing stone out in the middle of the field. We think of standing stones down in Stonehenge, but they are all over the place in Scotland. The stone was like a magnet. You go out there and put your hands on it, because you know everybody else did. Who put them there, when, who knows? But they're there. They have probably been there since the Earth cooled.[8]

Nobody was there. The substantial farmhouse was empty and appeared to be built about 1800. That was a point in time when small farms were being cut out of the large manors and they would put up a fairly decent house. Of course, this house was for the tenant. For the farmworkers, there were cottages all around and most of those were gone.

Right next to the farm is a highway, a former Roman road that goes from Perth up to Dunkeld. Along this road, there is a line of observation forts that runs east-northeast up to Perth and then turns north-northeast to a Roman fort that was big enough for a legion. (This had been Pict territory where native tribes would paint themselves blue and act the part.) This was the road that the Jacobites used when they came down off the Highlands on their way to the sea. Cromwell came up that road. The movement of people in Scottish history was up and down that road, within view of the farmhouse—history walking by. That changed my perspective on the farm and my ancestors.

The land was owned by General Graham, who served under Wellington in the Napoleonic Wars. I later headed down to the National Library of Scotland in Edinburgh and was soon holding the family's lease, a nineteen year agreement from 1798 through 1817. (Nineteen year leases for some reason were traditional.) At the end of the 1700s, with the agricultural revolution, many fewer people were needed to work those farms because of the changes in methods and the rotation of crops and pasture. The old common land was taken away. Many people headed for towns and cities to work in the weaving trades.

On the other side of the Tay River in a little community called Stanley, just two miles away, industrial revolution pio-

---

8. It turns out there is a line of standing stones that goes from Perth, which is the County Seat, up to Dunkeld, which is where the Tay River comes down out of the Highlands.

neer Richard Arkwright set up a water-powered cotton mill. This was the beginning of the Industrial Revolution and it was unusual, because Scotland had specialized in woolens. Why in the world would you want to weave cotton on the east side of the country when the only cotton you can get is coming in the west side ports? Moreover, instead of damming up the river, which would have been problematic for the salmon, they dug a tunnel from the downside of the rocks up to where the river was. The tunnel brought water for the mill. Did my ancestors go the short distance across the river to the mill to work? Perhaps. Here is a snapshot of the agricultural and industrial revolution within two miles of each other.

When the Napoleonic Wars ended, bad times arrived for the Dows. General Graham had doubled the rent. Prices were going up and the economy was in the tank. Soldiers were coming back home and didn't have jobs. There was a real question about what they were going to eat at night. Conditions got so bad that hungry peasants started attacking wagons carrying corn and rye down to Perth. There were food riots. In the midst of this, the Dows left Scotland and headed to America. Within six months, they had a hundred acres, and were off and rolling. It is my contention that the reason Scots came over to America was not the abstract notion of freedom, although I'm sure that played a part; the bottom line was owning land and the security that land ownership brought. That is the reason they went west and kept on going west. It was the land.[9] (It didn't matter that they were taking the land from Native Americans. That is another story.)

Today, the ancestral farm in Scotland is still operating, owned by an insurance company, which is holding it for development. Local farmers lease the land. There are some Dows in the neighborhood, but none of them claim to be related to me.

9. Editor's note: Wendell Berry recalls in *The Unsettling of America* a government hearing with then Secretary of Agriculture Earl Butz. When Butz mentions the option of American farmers leasing their land, a farmer stands up and famously says, "Our ancestors didn't come to America to lease land!"

I found an old map that shows the fields about 1816. The farm had a little more division then than now, but it appears to be growing the same crops today as in 1816: flax, rye, wheat, corn and turnips, and maybe even some potatoes.

And as you have now figured out, when I named the farm, I was still believing that the Dow clan came from the village of Ayrshire. It seemed appropriate then to name the farm after that special place. When I learned of our Perthshire origins, I had already picked the name and that left me with a choice: I could change the name or I could leave it just the way it was. I chose the latter as a little reminder to myself that you've got to do your homework. Don't get ahead of yourself. If you are not sure of something, say you don't know. The name is a daily reminder. Welcome to Ayrshire Farm!

Thinking back, if I had known what I know now about the family, growing up would have been a lot easier. I never knew any of the Dows. I would have felt more comfortable about myself, because there would have been some of them with whom I could have identified. Growing up, I used to think, why am I different from the rest of the folks? You don't want to be different; you want to be one of them.

(facing) A view of Bill's raised bed system, with remarkably clear paths and happy greens.

Growing

Conferring with a good bunch of young helpers before starting the workday.

Winter

It's the 29th day of February. Winter is almost over here at the farm. We need to start planting. Everybody on the crew is in gear and ready, but the temperature the last few nights has been in the twenties. So planting will commence probably the first of next week. This is early. These are always anxious times. You want to get the plants in the ground, but you don't want to do it so soon that you lose them. It is frustrating and a guessing game. I think we've got all the seed, but I may have forgotten something. We got the leeks straight to stick in the ground, but it's so doggone cold at night.

Winter for us usually starts the first or second week of January. At that point, we have finished the crops on the farm and probably our last deliveries. We have tasks squared away as far as the previous season is concerned. The next two months, January and February, are for fixing equipment and making plans about what we want to grow and how much and where. We are ordering seeds and plants, and making arrangements for transplants. We are getting in supplies like compost and feather meal—materials we are going to need, but won't have the time to go get when the time comes.

The thing about farm machinery like tractors and tillers is that they don't break unless you need them, and then you don't have enough time to get them repaired. If the malfunction is bad enough that you have to take it to somebody, you're in big trouble. You always *hope* that everything is fixed and ready to go. But you never know, and consequently, for us anyway, we've got two of everything as far as machinery is concerned. It is the best insurance you can have against being shut down because of a sudden loss of a machine. Here's my advice: Get two of your critical mechanical pieces.

Another thing that we work on in the winter is getting all the hand tools in shape. Handles get fixed; blades get sharpened. Maintenance is winter's work—big things and little things.

In the fall, everything has been turned in and planted with a cover crop. In our case, most of it is in crimson clover. We also use a plant called rape, which is a green. Rape does not have the nitrogen content that the clover does, but it goes through the winter. You can pick it and eat it, or pick it and sell it if there is enough. That works if the winter's cold weather hasn't done it all in.

We *could* say that the ground is at rest. It's not really the case though, because there is a lot going on in there. The clover comes up, and so does the henbit and chickweed. We started tilling the beds three weeks ago to try and kill off the henbit, and have them ready to plant. The growing areas are certainly calmer than during the spring, summer, and fall, but the soil is alive even in the dead of winter.

In the winter, three or maybe four weeks are spent traveling to Mississippi to see my brother and other folks down there. If finances will allow for it, we try to get overseas. I have a couple of family history projects I have been working on in England and Scotland and winter is the time to do on-site work on those. For me, winter is the time to just get out. I'm tired of farming. I just need to get away.

With the short days, I get a lot more reading done. I have books stacked high by my bedside and in the study—so high they are threatening to flood the rooms! Winter is not the prettiest time at the farm unless we've got a big snow out there. We haven't had one this year. It's cold. The dogs and I walk a couple of miles every morning. You hope that both you and the land are getting some rest. Rest can be lying down and sleeping, or it can be just changing what you are doing. That's the way it is for me anyway. The doing changes and there is some rest in that.

Tax time is coming. Doing the taxes is a long process because of all the farm stuff. There's separating out the bills: are these supplies; is this equipment; is this a repair? The upside of that process is that it's when we find out whether or not we made some money that year. We may have some money in the bank, but that doesn't mean a thing. I think about the fact that we just got through drilling a well. I don't know if I even have the money to pay for it!

I can't ignore that winter is a time of anxiety. Are we ready? Will enough workers show up when we need them? When do we put seeds and plants in the ground? It's a little too cold this week. Let's wait just a little longer. Since I have done it umpteen times before, it ought to be OK. I ought to have it right, but you just never know. Something could happen.

Drilling a Well

We have been trying to get a new well built since last fall. We've got to plant. And we need water to plant with. But it kept getting put off and that was partly my fault. The contractor finally got out here and drilled a well, and then the guys from the utility said you've also got to have a new transformer up the pole. It is going to be a lot of money out of my pocket, and by the way, they can't get one in here for six weeks. I can't believe in my wildest imagination that the utility, which two miles from this place has a maintenance yard full of transformers, poles and everything else, hasn't got a transformer over there that they can stick on one of these poles. You either wait or go over there and steal one. You're on their time!

Then you call an electrician about the well pump, because they have to come out and do this, that and the other thing, too. They are supposed to be here at nine o'clock, but you never see them and they don't call. Somebody else has to come in and bury the line, and they can't do it until the transformer is up there. It's a shell game and we're the ones left up against the wall. They don't seem to feel the same pressure. None of this is directed at me personally. I think that's where we are these days in our county. There's plenty of other work for contractors to do.

So we will start without a well. The trick to doing it, and we've done it before, is to plant right before you think it is going to rain. It's gambling, but you ain't got much choice.

Drought and Rain

Skip ahead to late May. With the weather having been a big issue last year, it has been great so far this year. We've had adequate rain and the timing of it has been good.

We certainly hope that it doesn't turn as bad as it did last year with the terrible drought. Dry conditions can help with some crops because they have fewer problems with disease. Tomatoes and peppers like dry conditions. But they've got to have some water. And last year there just wasn't any. If you are able to get the water to the plants (irrigation lines) and have an adequate source (a functioning well and pump), then they can thrive. Regardless, zero rain for months is a problem.

Other things that you grow in the summer don't like it so dry. Cucumbers are one example. And the planting of the fall crop and its production is really affected by them not getting rain in late August, first of September, and along the first of October. October is traditionally fairly dry and September we do usually get some rain, but nothing last year. Last year we also had a killing frost on Easter. So there were no blueberries, no apples, and no pears. That was a big issue for us.

For tomatoes and peppers, it was a good year. The only real problem we ran into was a couple of untimely rains. When you get a rain on them, tomatoes often times will split. You can have a good crop out there, and then you get a rain when it has been real dry. If the tomatoes aren't used to having that much water input, you can lose the whole set. In particular, the Sun-Gold cherry tomatoes split pretty easily, and there were times when the ground was covered with them. We had to pick them up and dump them. That hurts. But that's the way it goes.

In the spring, we need enough dry days to plant the summertime crop. It doesn't take a whole lot the way we are doing it. If we had to get into a field with a tractor, it would be a lot different. But our raised beds are very good at maintaining a constant moisture content. That is one of their real advantages. A lot of times we can get into that raised bed when the adjacent soil would be too wet or too dry to get into.

Now that it is May, we are finishing up the lettuce, arugula, mizuna, tatsoi, and rapini. The beets are starting to come out, too. We've still got a lot of leeks, but we pulled the last of the fennel the other day. The summertime plants are in and doing well. Tomatoes have really started to move in the last couple of days and the peppers are not too far behind. The basil that we transplanted looks pretty good, but the basil seeds haven't seeded in. The cucumbers are growing, although there is some disease in some of them. The blueberry crop looks very good, knock on wood. In two to three weeks, we ought to start harvesting blueberries.

This year we narrowly escaped the frost, which would have killed the blueberries two years in a row. On average, we get four out of five years without any particular damage as far as frost is concerned. We also have wineberries, not for production, but as more of a hobby. They are good. They require a lot of space and maintenance, which usually lends itself to chemicals to keep the grass down. In recent years, we have figured out a way to train them up the trellises, so they need less space and we can weed around the base more easily.

Crops and Cover Crops

In late spring, we plant a cover crop. A cover crop, for those who don't know that term, is a planting that covers and protects the soil and that can be tilled into the earth as a fertilizer. We usually plant edible soybeans and it takes ninety days to get them to maturity. We harvest the beans, sell those and then plow the root back into the ground. We have to get those planted in May.

Soybeans have pros and cons. The problem with the soybeans is that the deer love them. (Of course, the deer love clover, too.) So there is a competition as to who is going to win out. One of the things that makes the soybeans ideal, besides the nitrogen content, is that the timing is very good. We have a ninety-day window in the field and the soybeans fit that.

For fun each year, we try new varieties of lettuce, and this year has been no exception. (Now you know how market gardeners have fun!) A couple of new varieties have turned out very well: Galatia and Oxcard, both from Johnny's Seed Company. My simple notes: "nice color, nice appearance, good taste, grows well."

Another thing we try every year (and it doesn't always work out) is that we will put in another set of transplants along about the third week in March. If it works, we can have a very long lettuce crop.

This year the fennel crop came and went so doggone fast that we hardly realized it was there. We didn't take much to market—almost all of it went to restaurants. We have leeks left and radicchio. It is safe to say this was the best radicchio crop we ever had in terms of size and quality of the heads. You always lose some of them going to seed, but that is to be expected.

We grow a small crop of beets, and sometimes it works and sometimes it doesn't. This year the beets have been pretty good except for one thing. We have a very aggressive mole population and they like beets. We didn't realize the problem until we started pulling them out, and now it's a matter of trying to get rid of those buggers. They have certainly eaten their share.

The deer like beets a lot too and for some reason they haven't bothered much of anything else that had netting on it. We had netting on the beets, but our friends have been able to press it down and eat through the netting. Why they haven't attacked the other crops I don't know, and I'm not complaining.

This year arugula in particular has done very well, and we have sold lots and lots of it. Mizuna or Japanese mustard greens have done well, too, but the market for it is not as good as it has been the last couple of years. Our tatsoi or Japanese spinach is strictly for the foodie crowd; they love it and are willing to turn a blind eye to what it looks like.

Let me mention another new crop this year: cress. The way we got into it was that one of our helpers, Will Mitchell, had been taking a course up at the community college.[10] At the college they were growing something called wrinkled-crinkled cress. The seed catalog said that it was up to maturity in twenty-one days and I thought that must be a typo. By golly, not only will it do that, but it also grows well in the shade. We took some of the beds that are fairly close to the trees on the western side of the field, areas where nothing else will grow, and found that the cress is at home. Will and I are going to save those beds from now on for cress at $11 a pound. We can do that!

Do we grow crops that are special requests from buyers or just what we think will sell? We do both. There are several folks who buy from us who want certain things, and we talk about that before the season starts. I think of Nana's. Scotty Howell always wants fennel and we have gotten it to him. And Four Square. Shane has done a special with green romaine and we were able to keep up with him on that. At the Siena Hotel, Adam Rose has been great about buying a real regular mix of salad, lettuce and greens. Piedmont, Squids, 411—they have bought large, regular quantities of lettuce. And I should also mention regular customers like Margaret's, Lantern and Elaine's. I thank all of these restaurants and chefs.

Labor

In mid-May, the folks that we hire for the summertime have just been getting out of school and arriving at the farm. I usually don't hire that many students, but it is different every year. It turned out this year that those who were looking for

---

10. Central Carolina Community College in Pittsboro has an excellent sustainable farming training program.

something to do were students, which is OK in the respect that they are young and vigorous. They haven't had much experience, which isn't so good, and they go back to school the first of September, which really is not so good. I don't want to be disrespectful, but it can be like summer camp around here. I'm sure they can do all the computer stuff, but sometimes it's a matter of "which end of the shovel do I pick up?" OK, it's not that bad! But with the college helpers, sometimes I think they ought to pay me. I'm teaching them all day long. (Of course, most of them couldn't afford that.) About the time the workforce really gets up to speed and is of real value, they are going off somewhere. It seems to work out, but it does give me a little anxiety.

All that said, they are good kids. One young fellow may be able to stay around as a machinery man. There is a young lady who really knows how to work, and I like her. The others are fine, too. I'm going to get copies of Wendell Berry's *Country of Marriage* and give that to each one of them. I want them to know that it's not just food; *it's agriculture*. This is an early one, before *The Unsettling of America*. *The Country of Marriage* is one of those good ones. Check it out.

Our Latino farmhands, every one of them, are brave souls and great workers. They don't get off at the chicken plant until 3:00 P.M. If I'd been working at a chicken processing plant for eight hours, I don't know that I'd be too interested in going out and doing farm labor. Their work ethic is impressive. They are risking a lot, whether it be from getting a hand cut off at the processing plant or getting caught by the law out here in our fields. Every time a plane goes over, there is laughter and murmurs of "Immigration! Immigration!" It's not so funny, though.

We had three young women who worked here whom we referred to as the "pizza queens." They all worked at Pepper's Pizza. Now a pizza parlor may not sound like where you'd find your best workers, but they were great. When the weather was hot, they wore a minimum amount of clothing, which may have helped move things along in the summer. The lads certainly liked that! Actually, I think both sides enjoyed it.

It has been said that farming is not for everybody. But when I hear that I think: does that mean that being in physical shape is not for everybody and a proper diet is not for everybody and being outdoors is not for everybody? In a healthy society, everybody ought to be able to farm. Of course, they may not like the work, at first.

Folks have gotten a bit soft. We've lost our pride in being able to do physical labor. They used to have contests to see who could plow the straightest furrow or milk the cow the fastest. Similarly, folks have lost the ability to organize tasks on their own. We see it with children's games. If the little league isn't there to tell children what to do, can they set up and play a simple game?

Red Clay

On the farm, adequate water is essential. We can't do without it. During the big three-year drought, the well went dry and it was dicey. To deal with drought conditions, there are things farmers can do to help out. The raised beds are much more drought resistant than the old single row methods. Mulching helps. High quality soil, with enough organic matter in it, holds water well.

A lot of people look at the old red clay we've got around here and turn their nose up at it. But the old timers will tell you that if you've got red clay, it will hold water and you'll make it. I have reason to believe that they're right, because that's been the case for us. Now that doesn't mean that we haven't been dry sometimes, but we haven't been as dry as some of our neighbors. The old red clay gets mixed into everything else. It looks bad. It's not something that you pick up and it falls through your fingers like good loamy soil, but it's part of what you need.

As far as irrigation goes, we use low-pressure ooze tubes. It's made of ground up tires formed into a hose pipe. We've found those to be adequate for what we need. The overhead stuff is expensive and wasteful of water. Drip irrigation is appropriate for some things like blueberries but we haven't used very much of it. You've got to have water to begin with, but using it carefully is the key. And in the end, that is mostly about good soil.

In Ink

Good recordkeeping is vital. New farmers need to set up a system early on so they don't waste time trying to remember "what in the devil did I do last year, and the year before, and where was it planted, and how many bushels did it produce?" My record keeping for production, which is not all that it might be, is a listing of what we planted, including the variety, where we put it, what we added to the soil, and the dates of planting and harvest. I can go back and look at what happened for X number of years for the third row in terrace five. There will be notes in there about what did well, what didn't do so well, did this variety do better than that variety, and other comments. If you're not keeping those sorts of written records, you are causing yourself trouble. Of course, at my age I don't remember things very well, so I've got to write them down. Some folks may be able to remember everything for the last twenty years, but I never could pull that off.[11]

I also keep exact records on what we pick, what we sell, and what price we sell it for. I can tell you exactly what I sold to La Residence twenty years ago on this day or to Margaret's last month. That helps a lot when the chef says, "Well, we need X product, the same amount you brought us last week." I've got it written down and it's not in a computer. It's in a journal in ink.

When gas prices went up several years ago, I had to make a decision about selling to restaurants in Raleigh. I was supplying some very good restaurants over there and I enjoyed working with those folks, but was it worth the money in gas to go all

---

11. It always amazed me working with folks in the mountains, when I was up there in my youth. Some of the folks weren't necessarily good with reading and writing, but they could remember. They could tell you, "I bought a truck in 1960 for $18.73." You can take it to the bank—it was $18.73. Some people got it!

the way over there and back? After checking my records (and much deliberation), the answer was "no." Recordkeeping is key, and sometimes you even need to look back at them.

Machines

Ayrshire Farm is a small operation. With only five of its twenty-one acres cleared for agriculture and two of those in orchards, just three acres are in intense vegetable cultivation. At this small scale and using raised beds, hand tools are by far what get used the most. Tools such as hoes come in a lot of different shapes, sizes, and weights. We see certain tools being used before other ones, either because they are most appropriate for a job or folks like the way they work.

I like to have a number of different types of hand tools around for folks to experiment with. If we find one that's very good, we'll stock up. The farmhands will practically fight each other to get to use the diamond hoe—it's a high quality industrial hoe. Bottom line, don't use a tool that works you rather than you working it. You will know it when you use it.

In terms of maintenance, we make sure the tool edges are kept sharp and they're cleaned up each day. It's something that the farmhands need to know how to do. There's not that much to it—knocking the dirt off, applying oil and keeping an edge on it—and it makes a big difference in the ease of use and lifespan of the tools.

For raised beds on a commercial scale, a rotary plow is essential for us. There aren't enough hours in the day—not even close—to do raised beds without doing it mechanically. A rotary plow is an old design; it was used back in the Twenties and Thirties. It's a very good tool and will hoe, disk, and harrow all at the same time. It's got tremendous torque. I've been using Gravely's: they're good quality and heavy enough. They do good work.

I ask all the farm helpers to try their hand at the machines. This is not because I necessarily want all of them to use the

On a cool day, packing up some produce at the shed, with crucial equipment in the background.

equipment, but because I want them to have the experience and to see (and feel) what's going on. The first thing is to get them to realize that they can't out-muscle the machine. I don't care who you are, you haven't got enough muscle to out-do that plow. You've got to use your head and lots of leverage. That's the ball game. Once you've learned that, then most all things are possible.

We also use Troybilt 8 horsepower tillers. We don't do much tilling. After we throw up the bed with the Gravely, we come down over that with the Troybilt and flatten it out. That's all there is to it. If I had two more acres in cultivation, I might think about a small tractor. That's real money and we're not at a point where we need that. Any land in addition to what we're doing now, we'd have to go the tractor route to stay ahead of things.

With the big equipment, maintenance is a serious issue. We check the oil and grease whenever they get fired up. We change oil at regular intervals and keep good maintenance records. The log is hanging up there on the shed wall.

The Gym

Farming is physically demanding and there is no way around that. I grew up doing physical labor and I've told a lot of young folks this: If part of your self-concept is not wrapped up in your ability to do physical labor, farming may not be the thing for you. There are a lot of methods that make it not as laborious as it has been, but the bottom line is you still have to do the work.

When I think about how many people are over at the gym, it floors me. People pay to go to the gym to work out. I'm willing to pay *them* to come out here. Not only are they going to get a good workout, they get fresh sunlight, a tan, something good to eat and great conversation! Young people will go to the gym in a heartbeat, but the idea that they'll have to do physical labor out here is beyond the pale for a lot of them. They're not going to do it. It's too hard. It's too hot. Very unfortunate.

Yes, there are some bad days, and by that I mean, yes, it can get hot. But on a bad day we don't work between noon and 3 P.M. or even later. We stay out of the midday sun. We're just not going to do it.

On the other hand, sometimes it is good to learn what you can do that you didn't *think* you could do. You may be able to work harder and in tougher conditions than you first thought. And it is a good thing to have some appreciation for the people who do farm work all the time, not just a summer college gig. There are a lot of folks out there who do it all the time and all their lives. They need to be appreciated. A stint in farm labor is one of the ways of being able to identify with them and appreciating the dignity of their work.

One of the things I love about our farming operation is that we've got a mix of generations. We're in a place in our culture right now where old folks and young folks don't interact anymore. Has a young person worked shoulder to shoulder with an elder? Has an older person seen the latest dance craze or the latest clothes? Young folks learn from the older ones and the old folks stay young by working with the young folks. Farming is one of the places that that can be done.

I think about the older person who has some experience and knowledge. If that doesn't get passed along, then we are sentenced to repeat the learning of that same stuff over and over again. That's not the way to make progress. This is especially important in small-scale farming.

Plants

To grow plants, you need seeds. As far as where I get them, I am happy to say that I buy the vast bulk of my seeds from Johnny's. This is not paid advertising, but it is no secret that they've got good seed, much of it organic. They're easy to deal with. They're knowledgeable. They've made an effort to get out, see people, and talk to them. Some people are put off by the fact that the company is up in Maine, thinking that seeds from Maine aren't going to work around here. That's proven not to be the case.

These days our plants are really moving. Their color is good and they are really starting to look like something. They weren't all planted at the same time and we can really tell the ones that have been in there longer, how much more mature they are than the ones we just set out. The latter are just stalks with a couple of leaves on top.

When peppers and tomatoes start to set fruit, it is time to side dress. We add a soil amendment, in this case feather meal, about six inches around the plant and dig it in the ground. That will give it a little boost right about the time it is setting fruit, telling it to come on.

I grow organically, but I don't use companion planting on the farm. This practice of growing certain plants near crops to protect them from insects and for pollination is a good example of polyculture. Planting marigolds to control aphids on food crops is a classic example and for some people it works, especially in a home garden setting. It's not that I don't agree with it. The issue is that, in a commercial operation like this, it takes so much time to pick, to look through something and find one thing, that it isn't practical. It just takes too much time. Sometimes we do end up doing companion planting by mistake. (This can happen if certain weeds get into a bed.) Lots of good things can happen by mistake and we reap the benefits of our errors.

The thing that surprises everybody who is in it commercially is the huge amount of time spent picking. The trouble is that maintenance gets left behind, because it is more important to pick than do the maintenance work. The burden of weeds, the enemy, is always right behind you, shoving you on. Hurry up, get this done so you can go slay the bad guys!

What do I think about fungicides, the chemicals used against plant disease? There really aren't that many "good" chemicals, even if you could use them or wanted to use them. An organic farmer like me looks to other solutions. A prime remedy is hybridization—using plant varieties that are disease resistant. Another technique is crop rotation, in which we move crops around to let diseases and pests in the soil decline.

Rotation is very important—running away from the enemy, if you will. Because of the diseases in tomatoes and peppers, we are using a long six-year rotation. The tomatoes and peppers are not in the same place but once every six years. Even with that, you sometimes can't get away from certain diseases.[12]

Rotation of crops means that you've got to keep very good notes of what was where year to year. I sometimes think why didn't I keep better records ten years ago?! At the time, I thought I was doing OK with my records, but the detail turned out to be lacking. All you younger farmers out there, be sure to take adequate notes. I've heard too many novice farmers say, "Naw, I don't want to do that." What a big mistake that is.

As far as disease problems this year, something is going on with the cucumbers, which may be an insect or it might be a disease; we don't know yet. Time will tell. I believe that with good soil and vigorous plants, you don't get much disease in the first place. Of course, it is another matter to have experimental

---

12. This is especially true for plants that have diseases in the soil that are endemic, such as the wilt.

proof that that is true. One problem is that no chemical product is going to benefit from that research, so who is going to put up the money for that kind of experiment. A lot of the horticultural and soil research is being paid for by the CIBA-Geigys of this world.[13] So, do we get the research done that we need to have done? Who sets the research agenda? I don't know for sure, but we can guess the answer.

Soil Testing

I did some soil testing this year (I don't usually), and I did it mostly to satisfy the extension agent and folks in the soil lab. For some reason, it usually comes back about the same—it says that I need 100 pounds of nitrogen per acre. Several years ago, I asked somebody about that. "Why is it always the same 100 pounds of nitrogen? Does everybody else get the same advice?" He said, "That's because we can't really test for nitrogen, so we just assume everybody needs it." That assumption helps out one group in particular—the people putting the fertilizer in the bag. The chemical guys.

When I think about nutrients for my farm and ask myself if there is enough or not, I use the same principle, as in medicine. I go look at the patient. Get away from the tests and just go look at the patient. Are they doing better or are they not? Are they feeling better or are they not? That's what counts. I approach health the same way with the plants. If they don't feel so good, they don't look too good. Go look at the patient! And don't always start by running all of those tests. Money in medicine is in the testing and money in agriculture is in the chemicals justified by testing—that distorts everything. (Best doc I ever knew went to work for a private practice. After he had been there six months, they called him in and said—"You gotta

13. Editor's note: The Swiss company CIBA-Geigy merged with Sandoz Laboratories in 1996 to become Novartis, the world's largest pharmaceutical and agricultural chemical company. They have offices in North Carolina and around the world.

start doing more tests!" He said, "I don't need to do them." He wouldn't order them on anybody, and they were all over him.)

If I grow cover crops and turn them in, the data from a different piece of land in a different micro-climate says that I should have enough nitrogen in there. If the plants do well when we go to harvest, then I think they're right. If the plants didn't do well, then we need to do something different next year. What everybody wants is a uniform set of guidelines. That's not possible.

One day I called the soil lab and the technician said, "You know, your pH is up. I just can't understand that—you've got land that's traditionally very acidic." I said, "Yeah, but I've been farming it for twenty years and it works." "Well, I guess that's OK then." Of course, it's OK if I am seeing good results! But if I were a new farmer and maybe didn't know better or was intimidated by the scientists, then I might be out there trying to do something about "my problem" right away. That could be a big mistake. Sometimes I'm not very respectful of science.

Deer

This past fall we had a huge problem with deer. In spite of fencing, they ate the whole crop of peppers, which, having just done my taxes, I can say cost us somewhere in the area of $5000. That is a lot of money. Peppers are generally a good crop for us. When we lose the peppers and the cost of the inputs, it really hurts.

The deer started getting in when the peppers began to mature and flower. We got a few weeks out of them, but that was it. It didn't matter if we put the fence higher, left cars parked out there, or left radios on. It didn't make any difference. They wanted peppers. They broke down the fence or did whatever was necessary. October and November is when peppers really roll—they are one of the big crops for us in the fall along with tomatoes and basil holding over from summer. We like to end the year with a bang!

Because of the deer, the only bang we had was the sound of

a hunting rifle. To control the deer, we had several deprivation (hunting) permits. I could not guess how many deer were taken out of here. We worked with two guys who really knew what they were doing, but the deer just kept coming back for more!

Pretty Add-ons

Herbs do well. We sell them much cheaper than you can buy them at Whole Foods. As an add-on to the farm operation, they're worth it. They don't take up much space and are easy to grow.

As far as flowers go, there are several other local farms with great flower growers. We don't try to compete with that. What we've done is to find a few flowers that fit into our niche. Sunflowers are a good example as we can sell them to our restaurant customers. Another example is nasturtiums which make a great edible flower. The flowers, like the herbs, don't make much money, but every little bit helps.

Of course, the flowers do help attract beneficial insects, at least in in a minor way. That's good. We have many fewer bees than we used to and I'm not sure why. We used to keep bees, but I developed a bee sting allergy and we stopped. Like any organic farmer, I am *very* concerned about the future of our bees.

The Beds

Because of the limitations of our space, we are forced to grow on raised beds. We would do it even if we had more space, because it works, but here we haven't got any choice in the matter. The amount of produce we can get out of raised beds is far beyond what we could out of traditional single row farming. After many years of farming, my faith in the raised bed system is still strong. And the longer the bed, the better. I recently was talking to a colleague who was planning beds that were only about ten feet long. I said, "If you're really wanting to save space and really want to produce, you need long rows on flat, level

The parsley and other greens are really coming on, but so are the weeds!

ground. The time spent tilling 100 foot beds instead of ten foot beds is no difference."

I look at my field and try to calculate potential revenue per unit of space. Maybe I can grow great broccoli in that area, which might bring $2.50 per head. But I can grow leeks in that same space and make three times that much. I'm not the smartest guy in the world, but I can figure that one out. If both of them do well, then it's not a hard choice

We are growing on a slope with eight terraces. Say we've got a cover crop in there in the wintertime. That cover crop, which is usually crimson clover, needs to be turned in early enough in the winter or spring so that it deteriorates underground. Often, because of weather conditions, there's not a whole lot of time between tilling under the cover crop and when we plant. This year it's three weeks. We add feather meal (an organic fertilizer made from chicken feathers) and sometimes use wood ash, but not very much. Then we either direct seed or transplant into it.

We've got an old three-quarter-ton Ford truck up on leaf springs. It gets about five miles per gallon, so we don't fire it up very often and it never goes off the farm. Because it's up on leaf springs, it's up high enough that we can drive down a bed straddling the growing area and dumping stuff off the back—compost or organic fertilizer. This saves us a lot of time as opposed to doing the work by wheelbarrow load.

Except for some experimental crops, we grow the same types of produce every year. But within each crop type, we change the varieties a lot. In other words, we're doing variety trials all the time—of lettuce, in particular. Take green romaine. We'll put in three to four different varieties. We'll check out how it looks in the spring crop and how it looks in the fall crop. Then the seed company comes out with two new varieties the next year and we wonder if these might be even better. Just because it's good one year when you've got a lot of rain, doesn't mean it's good the next year when it's dry.

It's not hard science, because there are so many uncontrolled variables. A lot of the success is just luck. It hasn't got to do with being smart, hasn't got anything to do with being a better observer. It's just dumb luck. This year it worked; next year who knows?

Once the transplants are in there, they're watered, then they're on their own. For direct seeding, we usually use the broadcast method rather than rows. A lot of people like to plant in rows; it's more orderly I guess. To me it is wasting a lot of space. Most all our greens are done broadcast. People ask all the time, "Why don't you put those in rows?" I reply, "It's doing fine just the way it is." You don't have to look or measure a lot to know that there's more coming out of broadcast than out of rows. But everybody's got their own way of doing it.

When that crop is over with, it may have been weeded once or twice. Then, we go in with a big mower and mow it down, and turn that in. When the spring crop is at its summit point—if we've got enough time in there and if we can protect it—we'll sow edible soybeans. They're basically ninety days from the time you plant to the time you harvest and that's enough days to get you from the first of June to the end of August. That's then the time to plant the fall crop. So, if we're really on top of it, we can get three crops a year off that bed. That's pushing it. A lot of times that just doesn't happen, because we started too late or the weather wasn't right. The risk with the soybeans is that deer love them. They can wipe you out real quick.

The old folks used to plant by the signs and some farmers still do (biodynamics, for example.) I would certainly never say that that didn't work. I know a lot of people who believed in it, their life depended on it, and it worked for them. It could be fuzzy caterpillars or information from the Farmers' Almanac. I find that I have my hands full just dealing with the weather, let alone following the signs, too. I am fine with others using the signs as a tool. It's just not for me.

Balance

Let's talk pesticides and health risk. Take for example a pesticide like aldicarb (brand name Temik), used to fight root-knot nematodes. If there was good data that said that Temik used on crops was harmless to humans, proven through long-term tests with a large sample size; that would say something. But the testing isn't adequate. It's just not! I am not willing to put myself and my family and the people in my community at risk. I'm not going to sprinkle pesticides on my food at the table and I'm not going to add them at the farm. If the chemical was proven safe and prevented something that was really a problem for me, I might think about it. But it's going to be a long time before we have that kind of data. And not to mention the expense of all these chemicals; they ain't cheap. The organic way, until proven otherwise, is the way to do it.

At our farm, the only real pest problem we have is the deer. Insects pale in comparison. Some years we get some insect pressure; some years we don't. It may be the harlequin bugs one year, for example. None of this pressure is to the point where I would be tempted to use a chemical out of a bag. We don't need it. The deer on the other hand are terrible.

Why don't we have insect problems? I don't know for sure, but I think it's because of the diversity. If I were a bug looking for a place to land and eat, I'm looking for a monocrop. I don't want to spend my time wandering around looking for the next spot. I'm looking for that monocrop down there on that five acres that's breakfast, lunch, supper and a snack besides. At our farm, an insect has to hunt around and fight off predator insects, and we're changing crops all the time. It's just not very inviting.

Keeping the insect life in balance is a prime part of organic farming and I love to watch the natural balance at work. Take the tomato hornworm, for example. This insect, a large green caterpillar, likes to eat plants in the Solanaceae or nightshade family (tomatoes, potatoes, peppers and eggplant). Unchecked, it can do a lot of damage. But Mother Nature provides balance. In the absence of sprays, a hornworm is attacked by a little parasitic wasp that lays eggs in it. The caterpillar then ends up with the egg sacs on its back, which are easily seen, and it dies. This summer for example I saw only two tomato hornworms. That's it. And it's been that way for years. I didn't bring the wasps in here—they were just here. One of the things I tell the farmhands is this—do not under any circumstances kill a tomato hornworm. You are in deep trouble if you do. One or two hornworms can't eat that much and the wasps need them! Imagine if I go and spray the devil out of everything to kill off the hornworms, spending money and time when that wasp will take care of it for me. Bad idea!

Picking

Harvesting, or "picking" as I like to call it, is time-consuming and meticulous work. On the surface, skill-wise, it looks easy. Grab some greens or fruit and throw 'em in a basket. One of the difficult ideas to get across to new folks is that picking needs to be done in a very certain way. Newbies don't see it.

Let's say we are pulling fennel. Where you cut off that root is a very distinct line. Too high or too low and we've got a problem. To leave on too much of the root is unattractive, unnecessary, and wasteful. If you are selling by weight, that's not fair. It doesn't need to be on there. But we also don't want to short-change ourselves.

Over the years I've found that it takes a long time to get that knowledge across to my field hands. Is it because I'm not very good at explaining it? Maybe it is just tough to understand the finer points. At any rate, when you have to go through that same teaching process every year, it gets a little wearing.

As for the buyers, I like to sell them produce that has been picked and treated properly and that looks the way they expect it to appear. That being said, there is a difference between the expectations that have been groomed by the stores and what is reality. The stores want everything to be uniform. So you come along with a lettuce plant that is twice the size of what they are used to in the store and they don't know what to do with it. It's a great lettuce. That's the proper size for it, but it disrupts things. Produce that looks different won't make it into the stores or pass muster at the markets, which is sad because it is good food.[14]

A lot of folks harvest in the morning. Me, I don't like the feeling of going to bed at night, thinking that I've got to get up early in the morning to get the work done before the sun gets to it. Talk about a monkey on your back! We pick in the evening. We wait until after the shade has been on the crop for a good thirty minutes. (The farm's small scale and layout mean that a huge and friendly shadow forms on the fields as the sun passes west beyond the trees in mid-afternoon. The terraces run east-west and the woods north-south.) Once some of the field heat goes out of the plants, this becomes a much more pleasant time

14. The restaurants I work with around here, for the most part, know the difference, and are not deceived by appearances.

to work. You know that if you don't *quite* get it done, you've still got some time in the morning. Some of the pressure is off. A lot of people don't do it this way and that's fine with me.

We pick Friday night for a Saturday farmers' market. That means that our produce was probably ten to twelve hours post pick at market versus twenty-four hours post pick if it was picked in the morning. That's a big difference in my book.

Remember that a lot of produce does not hold up all that well. There are fruit and vegetable varieties that are very desirable, but don't maintain their appearance very long. That is the reason you don't see those grown commercially for supermarket sales. They can't handle that sales model.

Picking is like weeding in that it is potentially a real social activity. It is a whole lot easier to pick with someone to talk to, back and forth, then to just be there by yourself. Solo picking or solo weeding is a real monkey on your back. Now some things you've just got to do by yourself. I do those tasks myself. But I leave for the crew the things that can best be done as a social group.

We pick into a bag, a flat, a box, depending on what it is. After some clean-up at the shed, the product is ready to be put in the pick-up truck. Some items are easier picked in quantity and then sorted. Tomatoes, for instance, can be picked into a bucket, taken to the shed, screened, cleaned up, and checked for splits. That works much better than trying to make a decision in the field where you are moving all the time. Other kinds of produce such as blueberries don't need that kind of attention. We can just pick them straight into a flat.

Whether or not picking is backbreaking work depends on what you grow and what has to be done with it. Customers are always asking me why we don't grow strawberries. Now we do grow blueberries. So what's the difference? Blueberries we can pick standing up. Strawberries are right down there on the ground. It's a long, long way down there! You haven't got to be too smart to figure out which one you might rather do.

Some people want to do both and that's fine. But my back says otherwise.

One of the rarely mentioned advantages of raised beds is that it makes picking easier: the planting area is raised up maybe five or six inches. That is significant when we are bending over for hours at a go. For some valuable crops, there is no easy way to pick. If we are cutting lettuce or pulling radicchio, we're just going to have to bend over. That's the way it is. Hell, I wrenched my back the week before last just getting out of the truck.

To be clear, we are not picking at twelve o'clock on a hot July day. Most of our produce is picked in the shade and especially all our leafy stuff. Peppers and tomatoes can be picked in the sun, but I don't want anybody out there picking from about noon to about three o'clock in the afternoon. It's too hot. It's too humid. Go and sit down and take a rest.[15]

One of the negative stereotypes of farming is having to be in the hot sun in the middle of the day. But farmers and farmworkers don't have to be under the broiler. That is just the way it is done. It doesn't have to be that way. This is true for small farms and even some of the big ones. You can sequence your planting. You don't have to have that whole field come off at the same time. Among the commercial folks, there is a certain drive to get it all off in time so that it doesn't "linger in the field." What it does instead is linger around the warehouse.

As far as bad working conditions go, the moral thing is that you don't push that on anybody. You certainly don't do it to folks that you are employing in a family business. You know them. They're friends. Mistreating workers—in the end, it is going to come back and get you.

15. One of the nicest places on the farm is the little creek on the southwestern side. It's a great place to get cool on a sunny day when it's hot.

Moisture and Micro-Climates

I'm a farmer, so I talk a lot about the weather. This summer the situation was very different from last summer. We had rain on a consistent basis. As I've said, we need the water, but it promotes disease. As a result, we are probably at least twenty percent behind where we were last year financially. That's a big bite. (In addition, with the economy not doing well, folks have been ordering less.)

There are a lot of plant diseases that grow much better when there is a lot of moisture. When this happens, conventional farmers can pour on the chemicals and that can be effective for them. For us, one avenue is to pick varieties that are better adapted to the wet climate. We are grafting some of the traditional varieties that are susceptible to disease onto resistant root stock. Researchers at NC State are working on this and some of us have signed on to do field trials to see what happens. The difficulty with doing field trials around here is that there is so much variability. From here to town, you will go through four or five different micro-climates. And even right here, from the top of this hill to the bottom, it just ain't the same. It takes a while to learn all that.

These micro-climates are the very reason we have developed so many varieties over the years. The varieties are specific to certain places. What industrial food people have tried to do is homogenize. It is all Red Delicious. And that doesn't mean that Red Delicious isn't an OK apple. It is. But it's not adapted to my local micro-climate.

Imagine that my family had been farming this acreage for years, and my great-grandparents and great-great-grandparents started things out. Just by virtue of having been here they would have been experimenting all the time, by necessity. That constitutes a lot of knowledge! If for any reason the farm goes idle, then all they knew could get lost in a flash. The next farmer has to start all over again. And the resources to help someone start all over again are those that too often value the latest and greatest and most generic: The Extension Service. Big seed and chemical companies. The Agriculture Department.

About a month ago, we had a tropical storm with torrential rains. The storm washed out the driveway, necessitating an expensive repair. Even with our terraces, we got some erosion, but not too bad. I can only imagine what has happened and is happening to a conventional farmer here or in other countries with that kind of rain. Due to tree-cutting, plowing and poor farming practices, the farmland degrades until the family is forced to move on. You can follow the migratory patterns. The land wears out and people move.

Last year was bone dry and we loved it. We used our irrigation system to supplement when we needed it. When it is wet like this year, it is important to have a few dry days in a row to dry out the soil for planting. I went down to terrace number six to look at it earlier today. What I do is ball up the soil and see if it crumbles. It still packs. But tomorrow the soil should be just right.

Apples

Mother Nature has a way of correcting herself. Last year the apples were decimated. The drought and a late frost did us in. This year we have mountains of apples. Whew! While we had maybe four or five bushels of apples last year, this year we picked an astounding hundred bushels. We have never had apples like this. The quality, size and flavor have also been good. As a result, all my customers want apples and I am more than happy to sell them. In my experience, if you lose a couple of crops or even just one, it may overwhelm you the following year.

Here at the farm there are about thirty apple trees up here at the top, new ones, and fifteen or so older ones down in the bottom of the field. The ones at the bottom were planted about twenty years ago and are standard varieties. The ones up here at the top are from the heirloom apple expert Lee Calhoun and I'm just trying them out.

Now apples do have their challenges. An insect called the apple borer can be a real problem. No surprise, it gets up into the apple and bores out the center of it. One day several years ago, I was reading an article about a control method and I said to myself, "This could be good. We need to do this today." The method involves taking corrugated cardboard and cutting it into strips. With the corrugation running vertically, we then tie it very tightly around the tree. When an apple falls to the ground, the borer goes with it to spend part of its life span in the soil. When it is time to lay eggs, it climbs up into the apple tree and gets into the bark. The borer, poor guy, has a hard time figuring out between the cardboard and the bark. The method isn't 100 percent, but it sure makes a difference. Perhaps this is an example of how, on occasion, we can correct Mother Nature.

When this country was first settled, everybody brought apple trees with them. As time passed, farmers would carefully go through the apples on a particular tree and look for the best ones. They would take that seed and plant it and see what would happen. That's the reason there are (or were) thousands of varieties of apples. It may be they were looking for a pie apple or a cider apple or any number of uses, flavors, colors, textures or sizes.

I used to know the names of all of the apple trees here on the farm. I had it all nicely charted out on a piece of paper. I lost that dang chart. (I have torn this place apart twice looking for it.)[16] So I went out the other morning and filled yogurt containers with apples. I put a number on the bottom of the containers and made a new plan of the orchard. I took two whole flats of these apples up to Lee to see if he could identify them for me. I think the number of apples took him by surprise.

Lee and others have spent years and years looking for the old-time varieties. Initially, he was speaking to the extension people and then later on with people like me. He was talking to people and just asking them: Is there a tree around that is

16. Editor's note: Daryl did finally find it.

different from the others? Almost without exception all of the old-timers knew of a tree and were very proud to have anyone come out and see it. Lee would take grafts off of it and add it to his collection.

Lee has a farm which is almost like a research station.[17] He uses the esplay method which is when you flatten and train the tree using wires. His research is documented in a beautiful book, *Old Southern Apples* from UNC Press. Lee told me the other day that now he wants to rewrite the whole thing. They have found so many more apples to showcase.

As Lee is getting older, it is good to know that others are taking up his project. There is a young apple guy named David Vernon, who is up in Caswell County and is doing a great job. David has a much bigger apple producing operation. They have a huge barn up there that is made entirely of walnut. (Just think about that for a moment. A whole barn of walnut!) The service these guys have done and are doing is immeasurable. Of course, even though these guys have discovered hundreds of kinds of apples, most Americans still could name at most ten varieties, if they are lucky. And many of *their* grandparents probably have some of these old trees there in the yard. Might be worth a look!

Beauty

Moving from apples to something more abstract, let me say something about the aesthetics of the farm. I can't admit to putting a lot of effort into beautifying the farm, even though it is a beautiful place. On a farm, beauty comes in a lot of forms and mixes with the practical. Fields, forestland, paintings of fields, sculpture made with wood from the forest. Which is a higher art form? I think of a brightly-colored field of crimson clover. It's a beautiful crop and it serves a practical function.

---

17. Editor's note: Lee has sold all his apple trees to a father-son team, mentioned below, who transferred them northward to orchards at Century Farm in Reidsville, NC.

We have to plan the farm in such a way as to survive, but we can appreciate when plants bloom. It's a matter of getting your mindset into what's pretty and what's not, what's beautiful and what's not. Some don't see the beauty of the clover. Even though I'm planting it primarily because of its nitrogen, the sea of color is a welcome benefit.[18]

Dogs

There are three dogs on the farm right now and none of them are much help with the deer. But they are good dogs and play a real part in what's going on. Ching (Daryl's dog) is fourteen and has hip problems. Despite that, some days she just trots down past the blueberries and is like a new dog.

Katie is our golden mix and is a friend to everybody. She unfortunately does not know the difference between running through the garden and not running through the garden, but we forgive her. She came from Robin Kohanowich who runs the sustainable ag program over at the community college. (Katie had been abandoned and showed up at Robin's door.) She will chase squirrels, groundhogs, coons and possum.

Rudy is our arthritic black lab. At first, he was absolutely scared to death of everybody, but he has gradually come out of his shell. He has a great spirit and when we go for walks, he's going to be the first one to the top of the hill, no matter what. Katie could outrun Rudy every time, but Rudy is somehow always first. You can hear him coming, panting and struggling with his three-legged gait. What drives him, I ain't got a clue!

The farm is on a road named Friendly Pooch Lane. People would leave dogs and cats in green boxes, years ago, and a neighbor would bring them here. That's why we always seemed to have lots of dogs on the farm.

18. Editor's note: Bill once said, "Agriculture is the original form of art."

Winter, Part Two

In general, I like things on the dry side, weather-wise. This past fall it started raining and it has not let up. By Christmas time, everything was soaked and the wet has stayed. Nobody that I've run into has any memory of a winter like we have had. After a three-year drought, now we are too wet! As the old timers will say, it has been so wet that it will suck the boots right off your feet. It has been wet and cold—the water has been lying on top of the ground for about six inches and then it was all frozen under that. The top four or five inches would thaw out, and you'd just have this frozen muck.

In January, we had one week of fairly dry weather and got out there and planted fifty pounds of garlic. That is right much for the way we do it. The night after we planted, we had a huge rain and it just moved the garlic right on down the hill. Even with raised beds, we can't stand but so much rain. We lost the garlic and then it continued to rain and rain, and we even had a snow. The result is that now all we are doing is cutting firewood. It is simply too wet to get into the field. When the time does come to plant, I hope we are ready.

Ideally, at the end of February, we have some of the drier beds ready to go, to plant leeks, lettuce, all the greens, the easy stuff to do. It wouldn't take half a day. But the beds are just sopping wet. In my years of experience, this may be the worst winter weather ever. Nobody I have talked to can believe the amount of rain. (There always has to be a standard somewhere.) Two years ago we were begging for water. Well, we got it.

Too much water is more of a problem than too little. As long as we get enough rain that we can irrigate, we are good. But I can't dry out too much moisture. There's just nothing we can do about it. Our farm becomes a swamp. Now there is one exception to this: if we put up hoop houses and green houses, we can fend off the ill effects of too much water, somewhat. The water is still in the ground and there can be significant seepage underneath our structures, but we could get away with some practices that otherwise would not work.

On a positive note, firewood sells. The tatsoi has done well. And the cold weather may help with some pests, including fire ants. I just hope we can find enough work to hang onto our good employees. That worries me the most. Can they hang on with so little work?

We have had some bad times in the last few years. No doubt about it. But the thing about bad luck is that as you moan and groan about it, it's an easy topic to get together on. Who can tell the worst story? Shared pain brings people together or at least it can. And there has been some pain out there. This is just the way it works. It's always something. If it's not this, it's going to be something else. It's not usually this bad. That's the thing about farming. You take your licks, and you hope to make it financially.

(facing page) A bounteous display of the farm's early summer offerings including greens, leeks, kohlrabi, radicchio, fennel, dill, parsley, beets, and nasturtiums.

Selling

Delivering produce to Piedmont restaurant in downtown Durham.

Hail

I am associated with several other farmers through our joint CSA: Judy Lessler, Jerry Fowler and Stanley Hughes.[19] On several occasions, we had talked about hail. It is one of those things that is always in the back of your mind. It only needs to last for about five minutes and you have just lost a lot of money.

One day a storm was coming through, and there was a prediction that there could be some hail. I was over at Judy's farm, Harland Creek, a few miles from my place. I left there, and by the time I got to the main highway, the hail was coming down.

I made the mistake of turning to the right coming home as that was the direction from which the storm was coming. I thought it was going to break the windshield of the truck. Quarter size hail was just pounding. It went on for a good six, seven, eight minutes. The hail stopped within a half mile of my place, but Judy's farm really got it. It looked like someone had taken a bush hog and mowed the whole thing.

The good news is that the damage looks worse than it actually is. As long as the root is intact and there is some vegetation left on top, a plant will likely come back. It won't be the quality of what it was, but it isn't a complete loss. Plus there are always more crops to put in. So in terms of when it could have happened, it wasn't the worst time.

---

19. Editor's note: CSA stands for "community supported agriculture" and is a pay-ahead subscription service for allotments of farm products from one or more farms. The original intent was to provide upfront capital from the community to small farmers for a growing season, with the harvest split equally each week among those who bought shares.

For this type of damage, crop insurance might be able to help. It's expensive. A bigger potential risk is that the people who have been buying from you may need to find another source of produce. Losing a big customer could happen. This is an argument in favor of having a larger number of smaller customers, as with farmers' markets. Customers there are generally loyal, and if a disruption happens, will later remember that I have given them good food over the years.

As far as the effect on Judy, because there are other members of the CSA, everybody just antes up more and we go on. Folks pull together in a crisis.

Community

Our CSA is pretty simple. Like other community supported agriculture models, it is a subscription produce service. First, we figure out among us—the farmers, that is—what we've got that week that can go in a box that totals up to about the right amount of money and has a broad enough variety. Judy Lessler puts in some recipes that fit with the produce. For my part, I take fifteen boxes up to Chatham Marketplace and put them in the cooler. Customers come get their new box and leave their box from last week.

The majority of CSAs are single farms. Our model with several farms gives us more variety to put in the box. You hear of somebody who says, "I've had Swiss chard from my CSA for fifteen weeks. That's enough!" It's true—you don't want fifteen weeks of Swiss chard. A one-farm CSA may get in that kind of rut, but we are less likely to have that problem.

With a typical CSA, farmers receive money from their customers in advance, as with many subscription services. With our model, subscribers pay up front, but *we* don't get paid until there's a dollar figure put on what each of us puts in the box. It's calculated on the box being worth $15. The money we collect sits in a bank account and is there for an emergency. This is counter to the CSA model of using subscription funds to pay for

Working with Judy Lessler at her Harland Creek Farm, to pack boxes for community supported agriculture (CSA) customers.

growing this year's crop. With a group of farmers in the CSA, it is the only way we know to do it.

We had eighty-five subscribers last year. When we sat down in the fall to think about what we wanted to do this spring, the long drought was hanging heavy over us. We decided we had better cut back and backed off to what was supposed to be seventy-five. Several folks slipped in there, so we've actually got seventy-nine. The way things have been going this year, we could have done a lot more. Hindsight is 20/20.

Of those dozens of CSA subscribers, it's fair to say that I might recognize only a couple of them. This is nothing compared to my farmers' market customers. This year we had Field Days—one at Judy's and one over at Ayrshire—with required attendance. Can we get the CSA to be more than just pick up a box and check off your name? Folks are busy, but if we're going to create a real CSA, *community-supported* agriculture, folks have got to show up and see the farm, meet the farmer and the other subscribers. This builds community. Another chance to meet our CSA folks is when our blueberry pick-your-own season starts. I always hope that they'll be among the ones that come out to pick.

Many of my customers know something about me and that makes a difference. It really does. A lot of my customers have come out to the farm at one time or another due to the Carolina Farm Stewardship Association (CFSA) tour, which has gotten bigger and better every year. It's probably one of the most important things that CFSA does.

All this said, I don't know of anything that does a better job of bringing diverse folks together than the farmers' market. The market can be a great social leveler. We live in a world that's spinning away from us pretty fast, a world that doesn't have a lot of room for continuity and tradition. As human beings, we have a need for that. Not to sound like a Luddite, but what we need today is not going to come from talking on a cell phone or sending a text message.

Chefs

After a few years of selling at the Carrboro Farmers' Market, a fellow named Russell showed up one day and made me an offer. He was working at a restaurant on West Franklin Street along where Elaine's is today, not too far from where Pyewacket once was. He said, "Whatever you have left over from the market, bring it on over and I'll buy it from you. We'll serve it tonight." What better deal can you have? He would buy everything!

Russell was a good guy and in those days he sure helped me out. He helped me realize that I could successfully sell to restaurants, which was a confidence booster. Next thing I knew, David Bacon and Muskie Cates and all those guys at Pyewacket were wanting to buy broccoli. Then Margaret from Margaret's Cantina was willing to buy. Later, La Residence and Bill Neal got in there, too. Russell got that ball rolling.

I have sold to Phil Campbell at the Flying Burrito and the wonderful guys at 411 West and Squids (the Chapel Hill Restaurant Group). For a long time, Ben Barker at Magnolia Grill was

Discussing a produce order with Bret Jennings, farm-to-table supporter and chef/owner of Elaine's on Franklin, in Chapel Hill.

all about it over in Durham. There have been a lot of people—some newcomers, but a lot of them have been here a long time.

In the last couple of years, boy, there have been a lot of places in Durham. I don't know how many of these eateries are going to make it, but they're doing a damn good job to help the farmers and serve real food. This is certainly a whole new base. The Durham restaurant scene, plus the fact that the farmers' market is thriving over there now, means some good things are happening. It's certainly worth my while both financially and personally to go over there and deliver twice a week. There have been a couple of times I've had to make extra trips at night because somebody needed more of this or that. You say "OK" with a big gulp, because it's a long way over there from Pittsboro. But you do that. They've bought your produce and they've got a packed restaurant and it's on the menu! At that moment, they need some help. You just go do it.

I quit going to Raleigh a couple of years ago, as I mentioned before. I had been dealing with several restaurants, but the gas prices went up too high. I don't see the likelihood of being able to go back. My advice is if there is some geographic outlier on your restaurant list, I'd be careful to cut it. The eateries in Durham are all close together. The ones in Chapel Hill are in two concentrated areas.

Chef Adam Rose at the Siena Hotel was supposed to call me back last Monday about what he wanted, and for whatever reason, he didn't. Then he called last minute and said he really needed some items. If he says he really needs it, then I am going to get that product up to him. That is what you do for someone who supports you. I charged him ten bucks for the special delivery, which paid for the gas there and back.

Here is a typical springtime conversation with a busy chef. Short, but effective:

> "Hello, Michael? It's Bill. I suspect you are working hard. What can I do for you? Uh-huh. Yes, I have three-quarters of a box of small yellow tomatoes. They're OK if you are going to chop them or can be used as culls. OK, got it. I've already sold out of peppers. Sorry. We do have plenty of arugula, mizuna, and tatsoi. We also have these herbs: cilantro, basil, chives, garlic chives, rosemary, thyme, oregano, and lemon verbena. Great! We've got some lavender, plenty of apples and I might be able to get you a bucket of nasturtiums. Do you want an oyster bucket or yogurt container full? Did the nasturtium leaves work out? If we don't have enough flowers, we'll fill it out with the leaves. Thanks. Bye."

Gourmet

These days, "gourmet" is the name attached to some of the restaurants I supply. I'm not exactly sure what that means really. I know that we used to try and grow broad beans and I never could sell them very well. I kiddingly said to someone one day that they were *gourmet* broad beans and all of a sudden they were gone. My goodness! There is something about that word.

I'm not a restaurant critic. I don't know who is the best and the worst. There is no question that the whole approach to dining out, the quality of what people eat, and their expectations have gone way up over the last fifteen or twenty years. It's not the same scene. I have had the good fortune of dealing with some of the eateries that are considered to be at the top. Their chefs and owners are certainly nice folks and have been helpful to me. It has been a lot of fun to be a part of that. Whether all of them think of themselves as gourmet, who knows?

Chefs are looking for taste, appearance, reliability and price. Those are the big four. What do I grow that chefs get excited about? I would have to mention radicchio, the different varieties of lettuce, leeks, and fennel. Frankly, we don't grow it unless *somebody* gets excited about it. There are some things that they want that we don't grow, because there isn't room or because it doesn't work here. It's not like we at our small farm can grow every variety they want. But drawing from the many local farms in the area today, why pretty much anything the restaurants want is possible.

At the end of the year, I will talk to each of the chefs about what has happened over the last year. What could we do better? Any new products that they might want? It's an airing of what has gone on and what needs to change. We talk about quantities and how long the season is going to be for various crops. That last one is tough, because you both want the season to be as long as it can be, but we are at the mercy of the weather, pests and diseases. These are great conversations, with lots of ideas coming from the chefs, since, really, they are the food experts.

Cucumbers

Most years, I've had something of a lead person, a young laborer who is closely associated with the farm and who has been here for some time. My policy with these folks has been this: if they've got something that they want to try out—maybe an idea coming from a restaurant, maybe it's their own idea, maybe their brother said something to them—they should have the chance to try it. A good example of this was when Matt Jansen said that he wanted to grow cucumbers. I said, "Matt, I don't know that there's that much of a market for cukes. Could be though. If you want to grow them, go ahead and let's see what happens. If they sell, we'll put them in the rotation. If they make money, it's yours . . . in the first year!"

Over the years, we ended up with several good crops that came about using this method. I may have had my doubts, but sometimes the young folks have been right. The process has been fun, and it's important for the young farmhands. It gets them into studying both the cultivation of the new crop and also the marketing of it. The growing and the selling. These experiments give our young farmhands a real shot at doing something special.

Now, back to the cucumber example. Matt was able to grow us a good crop of cucumbers, and soon, like everyone else, we had a surplus of cucumbers in the peak season. Marketing all those cukes was going to be tough. Luckily, Daryl came up with a great idea. She figured out that we needed to give the shoppers cucumber samples in a more interesting way. How about a fresh cucumber soup? A cold Bulgarian cucumber soup. There was no cooking involved—just a blender and the cucumbers, a little bit of our garlic and dill, and some oil and yogurt. We made the soup fresh at the market and handed out samples to anybody who passed by. It worked like a charm. We sold a lot of cucumbers and the cucumber experiment was a success. In fact, we continued the soup tradition every year and many customers remembered it.

Life at Market

The number of vendors at the Carrboro Market was initially regulated by how many spots we could have under the roof of the shed. Later, we expanded that, with more and more folks selling from tents set up around the grounds. Henry Sparrow and I had been selling al fresco for fifteen years. Finally, four

Bill's usual spot at the Carrboro Farmers' Market selling the fall crop of apples, peppers and tomatoes.

or five years ago, two places came open under the shed and we grabbed them.

I liked selling next to Henry; he was a good guy. UNC President Bill Friday used to stop by and talk with him. They had both gone to Chapel Hill High, but Henry had a rural background, while Bill Friday was, well, Bill Friday. I wish that I had been smart enough to have a tape recorder with me, because the childhood stories they used to get into were priceless. Henry saw things from his hardscrabble perspective and Bill saw things from the perch of a university president. But they sure connected through those stories. I kick myself a million times for not having them on tape.

In the early years, the shoppers were older—more faculty and staff at UNC, hence people who needed to cook. As time moved forward and food became a big issue, we have seen more students and people recently out of school. It's no secret that the market serves in a big way as a social gathering. Some shoppers stay around for an hour. You see them going around and around, speaking to each other, catching up, "how are the kids?" etc. And it's a social occasion between the farmers and the people we're growing for. I've seen the kids grow up, parents die off, the generational movement. It's a community that wasn't there in the past.

A major change happened in the market when the Health Department said that vendors had to have a registered kitchen for prepared foods. Henry Sparrow's wife, Laura, made great apple pies. People would line up to get those fried apple pies. But she couldn't get her kitchen certified and that was sad. You have to have a stainless steel sink and other things. In my view, the regulations were set up more for a fast food restaurant. Who's got that kind of capital? All of a sudden, in one fell swoop, you cut out a lot of people who made a good contribution to the market.

Going to market on a regular basis lets me have frequent face-to-face interaction with the people who are buying. If you sell wholesale, you might *never* meet one of your customers. At the market, if I'm not there, a lot of people say, "Where were you last week?" That means something. For some of them, they are almost joyous seeing me every week. Other folks can be very serious, and I suspect there may be family problems going on, or maybe it's just too early in the morning. I hear lots of stories. There's a guy in the history department, and he and his wife are always traveling. They'll be gone for a couple of weeks, come back and have all these tales to tell. So, at times, I'm talking more than I'm selling.

I don't mind taking the time to talk for a while, but you've got to be careful about it. The trouble with talking is that if there is a group standing in front of your spot, they can shut you down for fifteen or even thirty minutes. They're taking up the space and nobody can get to you. We're talking about some serious dollars there. I've had to ask people to move along, and some don't much like hearing that.

A few years ago, at the height of the food safety worries about tomatoes, when across the country folks refused to buy tomatoes, people still bought lots from us, no questions asked.

It was a matter of knowing me, trusting me and trusting in our organic production. Relationships trump fear, when there is transparency and the trust that that brings.

The Pricing Secret

There is a perception that organic food from small farms is much more expensive than conventional, supermarket produce. I don't think that bears out as far as the farmers' market is concerned. To determine my pricing, one thing I do is watch prices in stores. I don't go to Whole Foods or Weaver Street to do that; I'm at Lowe's or Food Lion to see their price points. Those are the folks that I'm watching. And we can often beat their prices.

Back at the farm, our team discusses what a reasonable price is for this and that. It's informal. For market, I don't do complicated calculations of how much it cost me to grow something. That is way beyond my skill or desire. What I do though is to keep records of prices historically. I look at what things sold for last year. That's probably what it's going to sell for this year, give or take a little. "Let's see. I've had the same price for leeks for three years now at least. But the price of gas hasn't been the same. Let's go up a little." Then you try it out. If you've got leeks at 50 cents apiece and they don't sell, then you have to think about it. $2.50 a head for radicchio is what I can see being a very good price. Some weeks they all sell and some weeks they don't. That's the way it goes. It is certainly possible to have the price too low. And while gourmet and luxury items can fetch a premium, in the eyes of most consumers, leeks are leeks are leeks. Fennel is fennel is fennel.

Product and price competition can be tough. I have about one year after I introduce a popular new item before some other farmer comes in with the same thing. It used to be that I might have a couple of years before everybody else jumped on the bandwagon, but not now. One of the things that's always ticked me off is when other growers cut the price. Not only do they

With his beat-up truck, about to make a produce delivery.

copy what you're growing, but they cut the price on you, too. There are ethics involved in setting prices—especially when my ox is being gored! There is one vendor who would copy an item of mine and then cut the price in half. I used to get mad about that until I realized that we were both selling out. I won that one.

I don't know what the pricing secret is. If I've got something good and it's not selling, I think hmm, maybe we've gone a little too high on this thing. I don't walk around the market to see what prices are; I don't have the time. (I'm usually there by myself. I've got to get there early and get set up, and can't go off and leave my stand. By the time market is over with, I'm whipped.) There isn't price fixing. Nobody comes around and says, "Hey, beets are going to be a dollar and a half this week." That just doesn't happen. But I may ask one person, "What do you think such and such is selling for today?"

Is it better to sell by the item or the weight? It has to do with practicality. It takes a lot longer to weigh things. If I've got radicchio and not all of them are the same size, I might say, "Well, we're giving away that big one." But I'm OK with that.

(It's amazing to me that you can have big ones and little ones and somehow they all get gone, and it's not all the big ones that go first.) Mostly, I'm just struggling to keep up with the customer who has five different items and doing the math in my head, let alone weighing it. Now, if we had a bigger space and somebody else helping out, that would be another story. You can't keep people waiting when they have a dollar in their hand.

The Future of Selling

One difficult issue for me, I will be frank with you, is what to think about operations like ECO, Eastern Carolina Organics. ECO is an organic wholesale operation—they buy produce from farmers and market it to stores like Whole Foods and institutional buyers. They have a statewide presence and are growing rapidly. It's a co-op and the farmer owners pay twenty percent of their overhead to the folks running it for them. It's run by good people. They send trucks down east and out west and bring farm goods back to this area. My concern is that there is still a great need to set up local markets around the state and connect growers and eaters more widely. Wholesaling may make that more difficult if the best available organic produce is bought up by such operations.

It is true that the organic produce demand in the Triangle is outstripping the supply and a co-op like ECO makes some economic sense. My worry is what will happen when the price of fuel goes through the roof and a local supply of produce and local market links become essential for everybody. Talking to a lot of the old folks who lived through the Depression, they made it through that time by getting food from the local farmer. I hope we don't end up in another Depression, but I worry about our food security and the effect of wholesaling on our future.

In the long run, somebody needs to take the responsibility to help organize local markets in other places. There are restaurants in Beaufort, N.C., for instance, that are begging to have fresh, local and even organic produce. Let's make sure that the guy or gal doing the growing and the guy or gal doing the potential buying can get together.

I want to see direct marketing of organic produce move to new places and reach more people. It is possible. My fear, however, is that rather than starting new projects and establishing efforts in new places, everybody just wants to pile onto the existing work. The Carrboro Farmers' Market, for example, has been here for a long time and is successful, so everybody wants to sell at the Carrboro market. We don't need that. We need something as robust as the Carrboro market over in Sanford, down in Smithfield, all over down east, in South Carolina and Virginia. That's what needs to happen. Let's spread the word and spread the benefits!

Taking Stock

Talking about organic farming practices with an eager group of visitors.

Gatherings

Thirty years ago, one of the first things the organic farmers in the Carolinas did was to set up an association and hold an annual gathering. This was the genesis of the Carolina Farm Stewardship Association. The state's main agriculture school, NC State, was not at all supportive, with a few exceptions like Bob Miller, head of the Soils Department. So it was rather ironic that the first organic gathering was held on the NC State campus.[20]

If you go back and look at the conference agenda from the early days, it's always been about practical things: how to grow tomatoes or how to make compost. There were some esoteric discussions, too, but mostly we just needed to help each other with practical problems. This includes both marketing and production issues, because we didn't know what we were doing on both the growing and the selling side. Thank goodness CFSA was there in the early days!

The gatherings helped solidify a group that was strong enough to sponsor a farm tour in the Triangle. This was about twenty years ago. The two-day event invited consumers out to see our farms and was held each spring. These days about 3,000 people come out in the spring for the Piedmont Farm Tour. The farm tours have grown in size and number over the years, and

---

20. It's been said that with every great idea, first it is ignored, then it is ridiculed, then it is attacked and then finally it is wholeheartedly embraced and profited from. That is how I feel about our organic growing here on the farm. The folks at NC State and folks in the farm and business communities have come around, but it still sticks in my craw that we were treated so badly for so long.

are found in several cities, and in the fall and spring. That said, we should not forget there were a couple of times in the early days that the Triangle tour came close to being terminated.

The more people we can get out to the farm to see what's going on, the better off we all are. Of course, farmers on the tour need to make a real effort to explain what is going on—how to grow things without chemicals. It is not enough to just show people a pig or a sunflower and say "Hey, look at that." Get off your do-nothing stool and make it worth something! The farmers need to be available to take questions and be ready to be challenged. We know much more about farming than a typical visitor, so it's not like many people are going to stump us. And it doesn't hurt to occasionally say, "I don't know."

There have been times when a visitor and I have gotten into an argument. Peacefully, mind you. Folks come out and they are of a different mindset. Why they came at all, I'm not so sure. When that happens, I don't mind defending organics or small farms. In the past, people would fight me over organics and especially over whether you can make a living out of it. In recent years, that hasn't come up. It's not that the battle has been completely won, but folks can look at our farm here for thirty years, and it speaks for itself. I may not live in the local palace, but we have survived.

The people who are the most skeptical seem to be the ones who come from somewhere else. They may be from down east or up in the mountains, from South Carolina or Virginia. "The people where I live don't eat these kinds of veggies." Arugula. Kale. Tatsoi. That may be true, but you don't have to grow exactly this mix of produce. You can grow something else that does resonate with the people in your locale. Most people like to eat. They don't necessarily catch on to the latest fads—they don't have to. How about corn, beans, and tomatoes?

In the last couple of years during the tour, I've been aggressively up front about my support of Latinos. I've got no patience for people who are close-minded and selfish. It irks me for people in the supposedly richest country in the world to say, "Yeah, you're good enough to build a house for me, but you're not good enough to be a citizen." They clearly don't know their history, because there are very few families in the U.S. who can say that they were not immigrants in this country. Nobody invited them, I can tell you that. Certainly Native Americans didn't invite them. In fact, Native Americans tried their best to run them out of here! I have been known to launch into a sermon about that and dare anybody to take me on.

Losing the Farm

Despite the occasional feel-good story in the media, we don't have sufficient handing down of farming knowledge from parents to children anymore. The message from most farmers to their kids is "Get out." The Baby Boomer generation in farming is waning rapidly. The tragedy of it is that when they go, the land goes, too. They have convinced the kids that they don't want to farm. And the kids say, "Yep, you're right." Somebody comes along with a big pot of money and the farmland is gone.

Bottom-line, good agricultural land should not be used for housing. God only made so much of it and we aren't going to make anymore. In this area, any place you can grow a garden, any place flat—or even somewhat flat—ought to be off limits to the houses. We've got plenty of slopes. We've got plenty of places where there's too much rock or other impediments.

Take this part of Chatham County where our farm is: the beautiful farmland across the road is now all houses. Go down the road and across the creek. It's all houses. Both of those farms were very productive, excellent farms. Dave McCracken's place was the best farm in the county, way and above everybody else. But Dave told his three kids to get out of farming. They did, and it's houses now. Both he and his wife are dead. The kids are gone. The land is gone.

How does farming knowledge best get passed on? Everybody learns differently. I think it is the role of the school system to help young people understand how they learn best. Some learn through books and classes, and some learn by doing and need to get out there with a good mentor. In general, we have lost in a substantial way the mentorship by the old folks for the young. It used to be within the family, but today most of them just aren't doing it. This first crop of small, sustainable farms has done a good job of taking on young folks and mentoring them. I'm thinking of folks like me, Alex and Betsy Hitt, Cathy and Michael Jones, and Judy Lessler. We should celebrate and expand that kind of mentoring.

What is missing is the passing on of knowledge about the land over many generations. What happened on that land before we got it? What was good about it? What were the things to look out for? How to deal with the water? That's gone. When the history of a piece of farmland gets lost, we fall back into our old efforts of trial and error. We can't get that history back. I sometimes dream about having just an hour with the last several people that farmed this place. That would be priceless. Utterly priceless!

Now what would I ask them? Oh, there's all sorts of things. I see changes in the soil types from one side of our field to the other. What is that all about? What crops did well in certain places? Was there a fencerow here or there? Where was it that the tomatoes did the best? What were the seasons like? What varieties did the best? Tell me about your seed-saving practices and results. So much to learn! Nobody ever sat down and wrote it down; they just did it. They knew how to treat the land and the key crops. They knew when to run the plow and when not to run the plow. They were very knowledgeable and were the experts at a time when even the folks over in the universities deferred to them.

It is great to see that there are young folks over at Central Carolina Community College who are really eating it up in the books. And there are kids working on farms to gain experience. Both groups ought to be exposed to both ways of learning. There's nothing the matter with reading books; it's the only place you can get some things. On the other hand, to be able to

Chatting with organic pioneer and friend, Cathy Jones of Perry-winkle Farm.

turn over a spade of dirt out here and know when it's ready to be plowed—that doesn't come from a book. It's about a feeling and your senses—it's an art. So, I'm looking at a revolution in agriculture, bringing back the artisanship combined with good science.

GMO

The genetic modification of crops is a troubling topic of conversation, to say the least. Take for example what has happened to one of organic farmers' best friends, the naturally-occurring Bt toxin. Bacillus thuringiensis (Bt) is a bacteria that occurs naturally in the soil and can be introduced to farmland as well. Bt's importance is that it secretes a Bt *toxin*, in lots of different strains, and these toxins can be an effective control of many different types of pest insects. The toxin goes into the gut of the insect and kills it. Long story short, it is the best organic biological insect control, hands down. Organic farmers have benefited from this for years—until now.

The problem is that today seed companies are artificially inserting the gene for the Bt toxin into plant seeds. Plants can now produce the toxin themselves. While this might sound like a good idea, this gene manipulation is tailor-made to produce insects that are resistant to Bt toxin. Think about it. Acre after acre of a crop has this gene, encompassing millions of insects. It is only a matter of time before some insects develop resistance. Once the resistance happens, the insect population is off on a run again. This unfortunate development has already happened and is on the increase. We have got to be idiots to allow this, to do this!

As many people have heard, the large seed companies are using tough tactics against farmers that don't use their patented seeds. For example, seed companies have gotten to the point where they're hiring people to go out into people's fields to see if they've got any unauthorized GMO seed. Picture the seed saving (non-GMO) farmer who lives next door to a GMO operation. The pollen blows over the road and gets into the adjacent field. The company finds out and says, "We're going to sue you for everything you've got, because you've got some of our seed." Even if the farmer denies it, the big boys threaten a lawsuit and where's the money to fight that? The farmer, having heard of other farmers who lost court battles and ended up losing their farms, settles out of court and gives up saving his seed. What farmer can stand up against the likes of Monsanto? We're looking at some very difficult issues, the power of the big bucks and the result of weak political will. Where does all this take us? I guess we have to stay in the boat and find out.

The good news is that there are organic and non-GMO seed companies and seed projects doing great work.[21] As those projects grow and mature, our seed banks have got to be stabilized and properly funded. We have got to have strong seed banks

21. Editor's note: Notable examples include the Organic Seed Alliance (led by Micaela Colley), the Southern Exposure Seed Exchange (Ira Wallace), High Mowing Organic Seeds (Tom Stearns), Fedco Seeds (CR Lawn) and Sow True Seeds (Carol Koury).

for safety. Also, seed saving is a good practice, and one that I support. I must admit that personally I don't do it. For me, it's extra work to be done. There are just so many hours in the day. However, I'm very supportive of other folks doing it.

Corporate Ag

I am very much opposed to the buying up and patenting of seeds. I consider it a crime. That seed belongs to all of us. These corporations didn't make those seeds and they shouldn't be allowed to patent them. These days I get a seed catalog and look for the variety I had last year, and lo and behold, somebody has bought it and taken it off the market. That might be good for business, but it's not good for agriculture or society at large. And that's the place where they have got us. Fighting chemicals is hard enough, but going after seed patenting is going to be rougher. Seed patents are tough to break.

We are facing increasing reports of weed resistance to glyphosate (Round Up). It's what we should expect. It doesn't make any difference whether it's weeds or insects or diseases. When you grow with a monocrop system and you over-use a treatment, you get trouble. The same thing goes on in medicine. Why are our antibiotics beginning to fail us? They were used inappropriately for everybody who had a cold or sniffle, and now we've got all this resistance. Today somebody comes in with septicemia from staphylococcus, and they might die because of resistance to antibiotic treatment. The same thing is happening in agriculture. Nature is variety! When we ignore that fact, it is a pathway to disaster.

On the bright side, there are good institutions working on this problem, restoring balance to our growing methods. The Center for Environmental Farming Systems, led by Dr. Nancy Creamer, has done a great job. They are tackling these tough problems, which is exactly the sort of thing that the universities ought to be about. The university should not be about caving into corporate influence and only conducting corporate-funded research on more chemicals and band aids. The corporate folks are trying to corner the market and we can't let them. Keep them out of the universities—they haven't got any business there! I know, they *do* have business there, into the millions of dollars, but it's not to my liking. They shouldn't be there anymore than detail people should be in medical schools. But then, who is going to stop them?

I can see the corporate folks who run Big Ag coming after small-scale organic in a big way. "There's nothing the matter with this chemical fertilizer. On the other hand, that manure over there (organic, natural, untreated)—it's dangerous." Frankly, Corporate Ag scares the bejesus out of me: the takeover of seed saving is just plain scary. I'm sure the individuals who work in Corporate Ag, who talk about feeding the world and saving the environment, have souls and care about doing the right thing. But Corporate America itself, beholden to stockholders, has no soul and never will. That is no place to entrust our most precious resource: the ability to feed ourselves. That should remain in the hands of small, independent and ecological family farms.

I think people intuitively know that pop-tarts and gooey clusters are not good for them. While we need farmers to do all they can to grow healthy food for us to buy, we also need the medical profession pulling its weight, telling the community that healthy food *is* important. The medical and public health professions can educate us to care about where the food comes from and how it is grown. Don't just tell me to eat lettuce. What kind of lettuce? How was it produced? Was it grown organically? And where?

Hats Off

So, who are my heroes? A lot of people have influenced me, starting back with my great-great and great-grandfathers. They did what I'm doing, growing food without chemicals, but they did it back in the 1840s and 50s. It is an inspiration to me that it could be done. I even know the varieties of vegetables that they used. They wrote it down!

Wendell Berry is one of my heroes. We need the philosophical side of agriculture. It's a part of the story and it creates a much different profession for the farmer who is paying attention.

I love Debbie Roos, the small farm cooperative extension agent in Chatham County. Good lord, bringing her to our county was a good fit. The place was beginning to rock, and she showed up and we went into high gear. She has worked her tail off and with not much appreciation from the powers that be. I also love Nancy Creamer and her whole gang at CEFS. There may be similar people in other parts of the country, but Debbie and Nancy set the standard.

I take my hat off to everyone connected to the sustainable agriculture program at Central Carolina Community College and especially Robin Kohanowich, its director. In former days, when you thought about training in sustainable farming, UC-Davis, Cornell, and Iowa came to mind. Today CCCC is right up there with those schools nationally. I salute the people who had the ideas and sat through the endless meetings—and we sat and sat and talked and talked. The effort is paying off now. Not only does CCCC have an excellent farming program, but their new culinary arts program ties nicely into the small farm production.

I do not want to forget my other heroes—the folks who showed up at market on Saturday morning, and especially those who came consistently year after year. We would be no-where without them. As an example, I would mention again a person like UNC System President Bill Friday. It's not like he didn't have a few other things to do in his life. Yet he came and bought at the market faithfully, almost since it began.

I should mention some of the other people who inspired me, going back some years. They include Reese Roe, and also J. W. Bradley who was up in east Tennessee and was the leader behind the anti–strip mining group. He had more guts than anybody I've ever known. In the agricultural community, there was Hershel Liggon with the National Farmers Organization (NFO). Hugh Thompson was a Tennessee dairy farmer I got to know. The first time I went to his farm, there was a milk strike going on, due to the low prices. The strike was handled by the NFO, but the NFO's leadership wasn't as tough as some of the membership was, namely Hugh.

As far as heroes for the next generation, I look around at some of the current growers we have in this area. You can guess the names. We may see them up close and see all the blemishes, but history will see them as heroes. Let's celebrate them today. Never underestimate or overlook what is close at hand. We are blessed that there's a lot of them!

Success

Here we are in 2010. Next week I will be sixty-five, retirement age. I don't think farmers ever really retire. We just slow down bit by bit. There may be some who are glad it's over. But by and large, farming is not an occupation, like brick laying or punching a computer. It's a calling. It is a productive, learning, artistic experience.

To be honest, I don't know what happened to the last twenty years. I'm still forty-five in my mind. I'm serious. Having to sign up for Medicare was a real shock—it doesn't compute yet. I have a lot of things to do and time is flying by. That's the trouble with agriculture—I've got other projects that I want to do that I've put off because of the farm. All of a sudden I realize I'm sixty-five years old and I've got a lot of writing I want to do. It's probably not of importance to anybody but me, but I still want to get it done. Somebody snatched twenty years from me!

I have better days than others. Tax time is agony to get through. I'm not a very pleasant person to be around then. The rain gets to me after a while. It's like the heat in the summertime. Day after day after day. It starts to work on you. I just have to come to terms with it. Organic farming is about building the soil; things are always getting better and that leads to optimism. It's not like we're wearing the soil out (or wearing ourselves out). That can be hard to see in a year when the heavy rains seem to wash all that soil-building away, but I know it is true.

I don't regret being a farmer, far from it. If there's no agriculture, there's no civilization. The bubonic plague can sweep through, kill millions, and we can recover from that. If we inadvertently wipe out the ability to grow food, it's over.

I'm sure I could have done a lot of things differently. Sometimes it's personality, sometimes it's fate, and sometimes it's just plain stupidity. We could have made more money. But you know, money ain't everything. I would trade money for good health any day. And in general, the farming life is healthy and invigorating. Human beings are not geared to sit in front of a computer. That's not genetically what got us here. What got us here was our ability to work the soil, to grow things, to attend to business, to attend to life. Modern culture has gotten away from that. There's somebody to do this for us; there's somebody to do that for us. They call it the service economy and I don't much like it.

None of this is helped by Walmart sitting over there buying cheap goods made by Chinese workers in poor conditions. When I first moved to Pittsboro, just in that one block on Main Street, there was a great hardware store, a used clothing store and other useful retailers. I knew that if I broke a piece of machinery at a quarter to five, I could call the guy at the hardware store and ask him, "Could you hang on for just a minute?" He'd say, "Come on." Every time! He didn't have to, but he did. Today he's gone. The world we're getting in its place feels to me like an artificial, disconnected place.

It has been a long road. I count it as a success that we have been able to simply survive economically. We have been able to establish strong relationships with restaurants, something I never thought possible in the early days. We have pulled in a

Laughing it up with Daryl and farm-to-table chef, Angelina Koulizakis-Battiste, at the Farm to Fork Picnic, a benefit for the Center for Environmental Farming Systems.

lot of good people to participate in this movement. A lot of good people. Looking back, I am really surprised all this happened. Remember how humbly this all began. We've come a long way.

Here's an example of a payoff. There is a guy named Kelsey Seigel who had worked for me on the farm. Kelsey had come from California and was a good kid. After a few years, he headed back to California with a little more knowledge under his belt. Fast forward a few years and I was at the Slow Food Convivium in Turin, Italy. It was 2006. I was standing there with 5,000 other people in this huge amphitheater at the closing plenary of the Convivium. It was starting to quiet down. Suddenly I heard someone call out. "Bill! Bill Dow!" I turned around and thought, "Holy smokes, it's Kelsey Seigel." I was flabbergasted and embarrassed to say the least. As it turns out, he was running Alice Waters' Edible Schoolyard program in Berkeley. I felt very proud.

To farm, some people would say, is very foolhardy. Verging on stupid. "What the hell are you doing? You could be seeing patients!" But I love farming. I love that I can supply healthy food to lots of people. I love that I can mentor young people and be a good employer. I love that I can steward the water and the land. I love that I can grow things out of the soil.

That is what I stand on.

CPSIA information can be obtained
at www.ICGtesting.com
Printed in the USA
FFOW02n1743150116
20474FF

9 780997 043402